]

T_{his}

2

2

2

Minimalist
Mobile Robotics

A Colony-style
Architecture for an
Artificial Creature

Perspectives in Artificial Intelligence

Volume 5

Editor:

B. Chandrasekaran

Ohio State University
Columbus, Ohio

Minimalist Mobile Robotics

A Colony-style Architecture for an Artificial Creature

Jonathan H. Connell

IBM T. J. Watson Research Center
Hawthorne, New York

ACADEMIC PRESS, INC.
Harcourt Brace Jovanovich, Publishers
Boston   San Diego   New York
London   Sydney   Tokyo   Toronto

Copyright © 1990 by Academic Press, Inc.
All rights reserved.

ACADEMIC PRESS, INC.
1250 Sixth Avenue, San Diego, CA 92101

United Kingdom Edition published by
ACADEMIC PRESS LIMITED
24–28 Oval Road, London NW1 7DX

Library of Congress Cataloging-in-Publication Data

Connell, Jonathan H.
    Minimalist mobile robotics : a colony-style architecture for an
artificial creature / Jonathan H. Connell.
        p.   cm. — (Perspectives in artificial intelligence ; v. 5)
    Revision of author's thesis (Ph.D.—Massachusetts Institute of
Technology, 1989).
    Includes bibliographical references and index.
    ISBN 0-12-185230-X (alk. paper)
        1. Mobile robots. I. Title. II. Series: Perspectives in
    artificial intelligence ; vol. 5.
TJ211.415.C66   1990
629.8'92—dc20   90-37900
                                                                                              CIP

Printed in the United States of America
90 91 92 93   9 8 7 6 5 4 3 2 1

Editor's Note

Despite the fact that artificial intelligence (AI) lacks a unified methodology, until recently there has been at least one idea that was shared almost universally among all schools within AI: that intelligence is a process of manipulating representations of the world and ideas. Even more specifically, the representations were understood as discrete symbolic in character, i.e., symbols of the type used by Turing Machines, and the mechanisms of intelligence were assumed to be algorithmic.

Two distinct visions have motivated the AI enterprise, even though in practice most AI researchers didn't (and still generally don't) think of these two visions as potentially different: one idea, epitomized by the Turing test, is that of capturing the pure essence of intelligence as a disembodied representational system; the other, the idea of making a robot that sees, hears, talks, and perhaps even feels, and has a body, is captured by the R2-D2s and HALs of moviedom. But in fact, except for an occasional foray into the integrated robot world, AI has largely followed the Turing dream. The reason most AI researchers have not seen any inherent conflict in the two visions is because of the assumption that the way to build robots is to create a representation-processing thinking machine with sensors at one end and actuators at the other end. By this view, thought is disembodied. Sensors supply the representation of the world, and these representations are processed, resulting in additional representations that correspond to instructions to be sent to the actuators.

Interestingly, this "pure thought as symbol manipulation" view was applied not only to problems that are essentially symbolic manipulation in character, such as theorem proving, but also to problems that had to do with the robot's own body, such as planning movements of its body parts, or to phenomena in which sensory interaction with the world was very important. Some researchers have suspected that perhaps the problem was being made more, rather than less, complex by emphasizing the orthodox centralized symbol processing models. Attacks on the orthodox view have come from three directions:

i. Non-representational approaches.

Perhaps much of intelligent action does not require or use explicit representations and their processing. For example, a coin-sorter, such as the one that is used in most soda-dispensing machines, uses levers and slots that respond differently to the weights and diameters of various coins. This physical mechanism enables the sorter to identify the coins. A representational language is useful to *describe* the machine: a stage in the operation of the sorter can be understood by a statement such as, "At this point the system uses the weight to decide if the coin is a dime or a quarter." Representation thus may be a meta-language to talk about the phenomena involved rather than a literal occurrence. In perception, Gibson has long been associated with the idea of "direct" perception, a view that eschews representation processing in favor of direct mappings from sensory information to categorical perception, mappings that arise directly from the architecture of the system. Connectionism has been embraced warmly by many philosophers on the grounds that it provides such a nonrepresentational account of cognition. However, it can be argued that connectionism is as representational as the traditional symbolic systems, the major difference being the type of representation. (See B. Chandrasekaran, A. Goel and D. Allemang, 1989). Edelman has argued similarly that the neural matter should not be modeled as a representation processor but as something whose connectivity patterns get selected over trials: the neurons form connections, the successful connections are retained, and the unsuccessful ones stop contributing to decisions.

ii. Reactive approaches.

For many tasks, the appropriate architecture for producing solutions is one that is "reactive," i.e., the responses are indexed directly over the situation description, rather than resulting from complex problem solving using abstract world models. In real-world cognitive agents, in particular, the evolution of sensory apparatus is such that most actions are indexed directly by sensory abstractions. As actions are taken, the changes in the world are monitored directly and additional steps are taken reactively as well. A pioneer in this method of organizing robot action planning is Jappinen (1979), who built a system in my own laboratory for perception-directed skill learning that learned to navigate paths in a simulated world. The work of Agre and Chapman (1987) is a more recent example of an approach that uses perception-directed reactive actions as a way of responding to a complex environment without complex planning.

iii. Distributed approaches.

A third direction of attack involves some aspects of i and ii, but adds yet another twist. Not only may there be no need for complex symbolic processing on representations of world models, but the action generation may not be performed centrally at all. Brooks (1986) has articulated an approach for robot motion planning in which reactiveness of responses is combined with distribution of action-generation in a subsumption architecture.

Two books in the Series now arrive which show some of the best work in this new genre of AI research. Since they have interesting philosophical underpinnings in common, this Editor's Note serves as an introduction to both.

Jon Connell's book, *Minimalist Mobile Robotics: A Colony-style Architecture for an Artificial Creature,* is written in the direction described in iii: the robot has no central world models, but a set of distributed, local, partial models are coordinated in a subsumption architecture to achieve the robot's goals in the physical world. Randy Beer's book, *Intelligence as Adaptive Behavior: An Experiment in Computational Neuroethology,* abandons the traditional AI goal of simulating the highly symbolic linguistic and logical behaviors of human intelligence, and concentrates instead on understanding how simple nervous systems show complex adaptive behavior in dealing with a dynamic environment. His thesis is that high-level symbolic behavior should eventually be shown to be built on top of this adaptive organism, rather than as a completely separate logic or symbolic engine that merely monitors and controls the sensors and the body. Connell and Beer share a world view about the importance of moving away from the centralized model-manipulation paradigm of traditional AI.

The Perspectives in Artificial Intelligence Series sets for itself a goal of presenting works that point to or exploit interesting and provocative new directions in artificial intelligence. These two books eminently qualify under this criterion.

— B. Chandrasekaran

References

P.E. Agre and D.A. Chapman (1987). Pengi: an implementation of a theory of activity. In *Proceedings of the Sixth National Conference on Artificial Intelligence.* 268–272.

R. A. Brooks (1986). A robust layered control system for a mobile robot. *IEEE Journal of Robotics and Automation,* **RA-2/1,** March, 14–23.

B. Chandrasekaran, A. Goel and D. Allemang (1989). Connectionism and information processing abstractions: the message still counts more than the medium. *AI Magazine,* **9:4,** 24–34.

H. Jappinen (1979). A perception-based developmental skill acquisition system. Ph. D Dissertation, The Ohio State University.

Contents

Foreword

by Rodney A. Brooks

Real creatures operate in a world where no measurement is certain, no action is sure, and no belief is definite. The real world just happens to be such a place. How, then, can we build artificial creatures that can operate in these chaotic environments?

Classical AI has not tried to solve the whole problem since the days of Shakey at SRI in the late Sixties and early Seventies. Since then, researchers have worked on pieces of the problem. The central paradigm linking these pieces of work and partial solutions has been the idea that there should be a strong central model reflecting the outside world in some representational terms.

In this book Jon Connell presents a completed project that differs from this classical approach in two ways. First, he has built a physical machine which tries to solve the whole problem, albeit in a limited task space. Second, he has built a system that has no central models; indeed, it has only temporally transient and locally partial models, and it has no internal communication between procedures acting on those model fragments. The context for this work was building a mobile robot with an on-board arm that can navigate around unknown and unstructured office environments, and reach out with its gripper to collect empty soda cans, returning them to the location where the robot was first switched on. The robot is called Herbert.

By building a robot with a physical implementation of a distributed processor on board, Jon was forced to operate at a number of levels. The interactions between those levels sometimes became important in unexpected ways. For instance, as is often the case with physical media, it became clear that messages could easily get dropped in transit between processors. Since the bandwidth requirements on the wires was very low, rather than build elaborate hand shaking protocols which would have required more complex physical communications media, Jon's solution was to make digital messages approximate analog signals by continuously sending out values as long as they were valid. This solved the problem and also made the philosophy of using the world as its own best model even easier to implement. It made it more natural for sensors to feed into the network with the very latest readings from the world rather than succumb to the temptation of storing sensor values, or processed models. By removing the idea of an internally stored model, one also removes the problem of having to merge new data with old into a coherent picture of the world. This sounds like a coward's way out but, in fact, with noisy sensors this is often a difficult task. Jon's measurements of the characteristics of the particular sensors on board Herbert show just how difficult such integration might be. His distributed algorithms, however, have no trouble producing globally coherent behavior from this basis.

Jon further chose a non-traditional approach by having the intelligence of his robot consist of a collection of independent processes with absolutely no internal communication. There is a fixed priority arbitration network on their output actuator commands, but no sending of messages between processes. They all observe the world (including the robot itself) through sensor inputs and decide on an action that they want the robot to take. Furthermore, they do not retain any state corresponding to what has happened in the world for longer than a few seconds. Even with these restrictions, Jon has shown how a robot can appear to an observer to be successfully carrying out high level tasks, seemingly with goals and motivations, persistence and plans. In fact, as we read his description we find that the robot has no such entities internally. They are inventions of the observer. The robot carries out actions that appear purposeful and are, indeed, successful, but they are not built out of the traditional building blocks that we are all familiar with through our own introspection. Jon thus raises questions as to the worth of those introspections.

This book concerns an experiment in an extreme point in mechanism space. Absolutely no internal communication and no lasting state are radical design choices for a mobile robot. The lesson to be learned is not that these are the correct choices for all robots; indeed, often much power can be gained with just a little internal communication or just a little lasting state. Our naive impressions of what must be going on in an agent's head are not necessarily to be trusted. As we build complex systems to do complex tasks, we must always ask ourselves just why we are following a certain methodology. Is the maintenance of an accurate data base really fundamental to the success of a particular task? Should communications lines be drawn between all modules of a complex system? Or can more robust solutions be found by looping through the external world where the true situation always and aggressively asserts itself? Such questioning can lead us to engineering more balanced, more robust and better performing systems. These are the lessons to be learned from this book.

Lastly, it is worth noting the service Jon Connell has done for the mobile robot community by his attempts at providing data and analysis of actual robot runs. All too often, papers on intelligent and perceptual systems for mobile robots describe complex systems at great length and then have little evaluation of those systems. Often they say nothing more than that it works well. An outsider finds it virtually impossible to evaluate and compare such systems. Part of the problem is that the researchers themselves have no way of evaluating their own work. Jon has not solved that problem but he has taken a step in the right direction by collecting large amounts of raw data and collating it into a usable form. For example, it is possible to see how the arm can be opportunistic in its search for a soda can by examining different spatial traces of the hand annotated with an execution trace of which processes controlled the hand at what stages of its motion.

One of the good things about being a professor is the excitement of working with graduate students and learning from them. Jon Connell had a plethora of good ideas, and has put them together well. I've learned a lot. Thanks Jon!

Rodney A. Brooks
Massachusetts Institute of Technology

Preface

This report describes a distributed control system for a mobile robot that operates in an unmodified office environment occupied by moving people. The robot's controller is composed of over 40 separate processes that run on a loosely connected network of 24 processors. Together this ensemble helps the robot locate empty soda cans, collect them with its arm, and bring them back home. A multi-agent system such as this has many advantages over classic monolithic controllers. For instance, it can be developed in stages, each new layer building on the last. It can also be split among several processors or agents so, as new capabilities are required, new hardware can easily be added. Furthermore, its performance degrades gracefully – if a single agent fails, the robot continues to function, albeit at a lower level of competence.

However, to achieve these goals, the system must be decomposed according to certain guidelines. First, the internal workings of each agent should be isolated from all other agents. This improves the modularity of the system and helps prevent implementation dependencies. Second, all decisions should be based on spatially and temporally local information. This keeps the robot from relying on incorrect models of its dynamically changing world and allows it to operate with incomplete sensory input. Unfortunately, these restrictions make it nearly impossible to use conventional techniques to perform tasks requiring spatial reasoning. The can collection task is particularly difficult because it requires three different types of spatial knowledge. The robot must be able to navigate through its environment, recognize the shape of a can, and determine how to move its arm for grasping. To build a functional robot we had to develop new ways of thinking about these operations. The rest of this report details the development of suitable strategies, discusses principles for achieving a satisfactory task decomposition, and examines the limitations of such a system.

I would like to take this opportunity to thank Rod Brooks for establishing a fine research laboratory, generating interesting ideas, and giving me free rein on this project. Thanks also to the other members of my committee, Tomás Lozano-Pérez and Marvin Minsky, for their invaluable comments on the thesis itself. I am indebted to the other people in the mobile robot lab, particularly Anita Flynn, Paul Viola, and Mike Ciholas, for a good environment and invigorating discussions. Much of the required wiring was performed by a series of undergraduates, without whom Herbert would be just a box full of parts. Special thanks to Peter Ning, who helped develop and construct most of Herbert's systems, and who put up with an inordinate amount of griping along the way.

This report is based on a thesis submitted in partial fulfillment of the requirements for the degree of Doctor of Philosophy in the Department of Electrical Engineering and Computer Science at the Massachusetts Institute of Technology in September 1989. This report describes research done at the Artificial Intelligence Laboratory of the Massachusetts Institute of Technology. Support for this research is provided in part by the University Research Initiative under Office of Naval Research contract N00014-86-K-0685, in part by a grant from the Systems Development Foundation, and in part by the Advanced Research Projects Agency under Office of Naval Research contract N00014-85-K-0124. Mr. Connell also received support from a General Motors graduate fellowship.

Chapter 1

Introduction

In this report we describe a real, fully-functional mobile robot which operates in an unstructured environment. The robot, called Herbert, is 18 inches in diameter and stands about 4 feet tall (see figure 1-1). It has a three-wheel drive configuration which allows it to turn in place and go forward and back. There are two rings of 16 infrared proximity sensors on the body for obstacle avoidance, and a flux-gate compass for navigation. To provide more resources to control, there is also a 2 degree of freedom arm on board with a parallel jaw gripper and a variety of local sensors. To provide a richer sensory input, the robot also has a high resolution laser range finder for locating and recognizing objects. The robot is completely autonomous and has all batteries and computers located on board.

Herbert has a single task: to collect empty soda cans. It starts by wandering around its environment and searching for cans with its laser light striper. When it finds a promising candidate, it carefully approaches and aligns itself with the target. Next, the robot releases its arm which gropes around using local sensors and retrieves the can from wherever it is sitting. When the robot has retrieved the can, it slowly navigates back to its home position and deposits its trophy. Finally, the cycle repeats and the robot ventures forth once again on its quest for cans.

The construction of this robot required us to integrate the spatial reasoning problems of recognition, manipulation, and navigation into a complete, operational system. This endeavor was greatly simplified by the use of a novel distributed control system. Instead of having a centralized sequential program, the robot is controlled by a large collection of independent behaviors. Each of these behaviors contains some grain of expertise concerning the collection task and cooperates with the others to achieve its task.

1.1 The task

The primary goal of our robot is to collect cans. This task was chosen partly because cans are such easy objects to identify. They are all the same size, rotationally symmetric, and typically found in a vertical orientation. This is important because it allows us to separate the variability of the environment from the variability of the grasped object. By simplifying the task of object identification we can concentrate instead on the difficult problem of moving through an unknown, cluttered workspace. Another reason for chosing this task is that it can easily be mapped to a number of different applications. For instance, instead of finding cans in a laboratory, the robot might be collecting rocks on the surface of Mars. In a more domestic capacity, a similar robot might be used to clean up the floor of a child's room. The cognitive and motor skills involved in all these tasks are similar.

Still, before extending our system to other domains, we must solve at least one instance of the basic problem. The usual

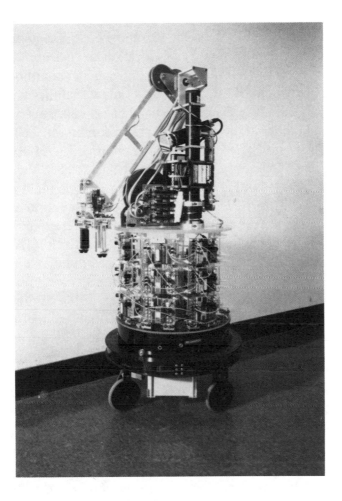

Figure 1-1. This mobile robot is named Herbert. Its only goal is to collect empty soda cans. It locates them with a laser light-striper, grabs them with a 2 degree of freedom arm, and brings them back using infrared proximity sensors and a compass for navigation.

approach would be to start by describing our target can as a cylinder of certain dimensions. We would then use some sort of remote sensing coupled with a recognition algorithm (e.g. [Grimson and Lozano-Pérez 84; Lozano-Pérez, Grimson, and White 87]) to determine the object's location and orientation. We can not just enter the relevant parameters via a keyboard because the actual placement of the can is part of the variability we wish to retain. After finding the can, we would next construct a three-dimensional model of the environment surrounding it. To do this we could use any of a number of techniques available for acquiring range images (e.g. [Echigo and Yachida 85; Vuylsteke and Oosterlinck 86]). However, all these methods yield only 2 1/2 D sketches [Marr 82]. To build a proper three dimensional model, we need to take multiple images from different directions and then fuse them into a single coherent description [Ayache and Faugeras 87; Porrill 88]. Once this was done, we would employ a path finding algorithm to plan a trajectory from the start point to our target (e.g. [Lozano-Pérez 86]) and then pass this path to a servo system to give us precise control of the robot's joints.

Unfortunately, the approach described above relies on extensive, accurate world modelling. The sophisticated sensory capabilities needed to support this endeavor are typically both expensive and difficult to implement. Also, many of the best sensing techniques require special lighting or impose restrictions on the surface properties of objects. This precludes their use in general, unstructured environments. The situation is further complicated by the fact that sensors mounted on a moving vehicle are unlikely to remain calibrated for any length of time. Furthermore, even if we could obtain clean data, the sensor fusion techniques necessary for building solid models are still under development. Finally, we are still left with the problem of bringing the can back to a central repository. Even the most advanced navigation systems (e.g. [Moravec and Elfes 85;

Chatila and Laumond 85]) make heavy use of world modelling. Thus they are fraught with the same difficulties as the identification and manipulation phases.

We believe the difficulty with the traditional approach stems from the centralized nature of world models and path planners. We are forced to carefully funnel all the sensory inputs into a highly distilled form and then use this "snapshot" to plan an essentially ballistic trajectory for the arm. To overcome this limitation we follow Brooks [Brooks 86] and adopt a "vertical" decomposition of our control system. Instead of having a single chain of information flow - perception system linked to modelling system linked to planning system linked to execution monitor - we have multiple paths. Each of these paths is concerned only with a certain subtask of the robot's overall operation, such as avoiding collisions, searching for cans, and achieving a firm grasp. We refer to such paths as "agents" and designate the specific functions they perform "behaviors".

The advantage of having multiple parallel control paths is that the perceptual burden is distributed. Within each path we still have perception, modelling, planning, and execution. But these components do not have to be general purpose; they only have to pay attention to those factors relevant to their particular task. For example, if the task is avoiding walls, then the associated control system does not need to know the color of the wall, its texture, its decomposition into generalized cylinders, etc. Similarly, we adopt a "reactive" style of control in which each commanded motion is based solely on the robot's current sensory information and goals. For instance, if an obstacle is encountered we simply instruct the robot to proceed parallel to the surface of the obstruction until it vanishes. Thus, we do not have to make a complete plan at the outset, but can instead improvise as we go along.

1.2 Animal stories

As a contrast to the usual robotics approach, let us examine some work from the field of ethology, the study of animal behavior. Much effort has been devoted to finding the "releasing" stimulus for particular behavioral pattern. By carefully controlled studies researchers are able to determine what features of the situation an animal is paying attention to. Typically, creatures do not have very detailed models of the objects they interact with.

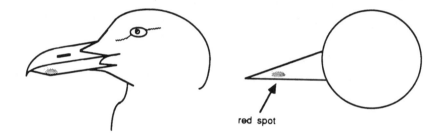

Figure 1-2. Animals seem to use incomplete models for many activities. Baby seagulls respond just as well to the mockup on the right as they do to their own parent (left). The critical features are that the object must be pointed and must have a red spot.

For instance, when baby seagulls detect the arrival of one of the parents, they raise their heads, open their mouths, and start squeaking in a plea for food. However, the baby birds do not recognize their parents as individuals, nor are they good at distinguishing seagulls from other animals or even inanimate objects [Tinbergen 51]. As shown in figure 1-2, the birds respond as well to a very simple mockup as to the real parent. The important condition seems to be the presence of a pointed object with a red spot near its tip. In their natural environment, this model works just fine because the real parents are the only objects which fit the bill. The same sort of minimal representations have been discovered for many other animals as well. This suggests that we might be able to build reasonably competent mobile robots without investing a large amount of effort into building detailed representations.

Another, more complex example involves the coastal snail, *Littorina*. In analyzing a behavior pattern researchers often tease it apart into components and describe it in terms of a number of competing urges or "drives". For instance, the coastal snail has some behaviors which orient it with respect to gravity and other behaviors which control its reactions to light [Fraenkel 80]. Sometimes these behaviors are in conflict and the creature is forced to choose one over the other. The actual behaviors used, and the manner in which they are switched in and out, can give rise to some interesting emergent properties. In fact, the snail can perform some seemingly sophisticated navigational tasks with a relatively simple control structure. Like Simon's metaphorical ant [Simon 69] the complexity of a creature's action are not necessarily due to deep cognitive introspection, but rather to the complexity of the environment it lives in. Our robot, Herbert, is built to take advantage of this sort of phenomenon and is named in honor of the originator of the idea.

But let us get back to the particulars of the snail. Figure 1-3 shows the creature's entire repertoire of behaviors and summarizes the postulated interactions between them. Snails have two basic reflexive behaviors which we will refer to as UP and DARK. UP tells the snail to always crawl against the gravitational gradient while DARK tells it to avoid light by crawling directly away from the source. It should be noted that neither of these "instincts" are complete functions: there are some input configurations for which they do not generate an output command. For instance, if there is no appreciable intensity difference between directions, the DARK behavior is quiescent and the snail crawls straight upward. Similarly, when the snail is on a more or less flat surface, UP is inactive and the snail's direction of travel is determined solely by the illumination gradient. Overall, however, DARK is the stronger behavior. If a very bright light source is present, the snail will crawl away from it even if this means going downward. In general, however, the commands from the two behaviors are combined and the animal crawls at a compromise angle. In figure 1-3 the interaction between DARK and UP is shown by suggesting that the output of DARK replaces the output of UP (circle with arrow entering it). However, this diagram is merely intended as a schematic representation of the interaction. The animal's nervous system might achieve this merger in some totally different way.

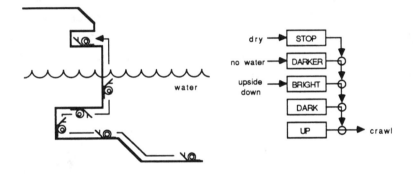

Figure 1-3. The coastal snail may be controlled by a fixed hierarchy of behaviors. The combined effects of these behaviors enables the snail to navigate to its feeding area.

Surprisingly enough, if one turns the snail upside down, instead of avoiding light, it will now head toward bright areas. We can imagine that this is due to a third behavior, BRIGHT, which provides the animal with a positive phototaxis. Since BRIGHT ends up controlling the motion of the animal, it must override the output of DARK. Yet this new behavior only becomes active, "potentiated", when the animal is inverted. Otherwise the creature acts solely on the basis of the lower level behaviors. This is an example of a behavior which is more than a situation-action type reflex. It is a control pattern that is switched on in reaction to environmental conditions beyond those necessary for the orientation response. It has been observed, however, that this light seeking behavior occurs only underwater. If the animal is in air it will invariably seek out dark areas, even if it is upside down. This can be modelled by adding yet another behavior, DARKER, which, like BRIGHT, is only active in certain situations. When the creature is out of the water, this behavior takes precedent over all the other light sensitive behaviors. The actual reflex response embedded in this behavior is nearly identical to the one in DARK but somewhat stronger. At high illumination levels DARKER will halt the snail in place, whereas on a sufficiently vertical slope DARK would have allowed the animal to crawl toward the light.

The last behavior, STOP, halts the snail when it encounters a dry surface and thus keeps it from wandering too far inland. Unlike the other behaviors discussed, it has a completely specified reflex component. Yet it is of a special type, called a "fixed action pattern". This function does not depend on any sensory input at all, it simply produces the same output every time. This behavior's potentiation, on the other hand, does depend on sensory stimuli. In this case the halt response is only evoked when the snail fails to detect dampness underneath. Our catalog of behavioral primitives for modelling the operation of an animal is now complete. In our descriptive "language" the motor control signals for an animal are generated by the reflex component of a behavior and can either be fixed or vary based on sensory information. In addition, some behaviors also have a gating component which activates the reflex only under certain conditions. Finally, the results of all the behaviors are combined, often by exclusive selection, to produce the actual motor command.

Fraenkel explains how this collection of behaviors aids the creature in its pursuit of food. These particular snails eat algae which grows in the cracks between rocks slightly above the tideline. The behaviors we have discussed help the snail reach this food source and prevent it from being cooked in the sun. Imagine, as shown in the left of figure 1-3, that the snail starts off on the ocean floor a short distance off shore. Since the rocks are slightly darker than the surrounding sand, it crawls along the bottom towards them. When it reaches an outcropping it starts climbing the face. If it comes across a notch in the rock it is first drawn inward by negative phototaxis. Upon reaching the end, it then starts climbing the rear wall and eventually reaches the ceiling. Here, it becomes inverted and thus moves outward toward light again. Having successfully overcome this impediment, the snail continues climbing toward the surface. When it reaches the edge of the water, if the sun is too bright, it stops and waits. Otherwise, it ascends still further until it reaches a dry area or comes across a crack. As before, the dark seeking behavior will take over and directs the snail into any crack encountered. However, since it is now above water, the snail does not turn around when it reaches the back, but instead stays deep in the crack. It presumably stays there until a wave washes it back into the sand.

The snail thus arrives at the region of maximum algae concentration even if it has to negotiate major obstacles along the way. Yet it does this without any conscious knowledge of its purpose. It has no understanding that it is trying to reach the first crack above the waterline, it merely acts as it was built to. We have designed our robot, Herbert, based on this same principle. As with other "creatures", he can not be told what to do, he simply behaves according to its nature. If we want Herbert to do something else, like deliver mail, we would build a different robot. Fortunately, many of the underlying behaviors necessary for these two tasks are similar so we would not have to start from scratch. Still, there is no way to directly tell the robot which soda can you want it to pick up or where you want it placed. The best we can do is start the robot off near the desired return site.

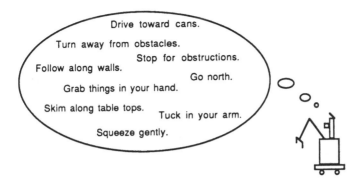

Figure 1-4. In our colony architecture there are a large number of independent control units which all operate in parallel. Each of these agents is responsible for some small part of the robot's task and compete to control the robot's actuators.

1.3 Design principles

Our approach to robot control is derived from the ethological perspective. We visualize a whole colony of locally autonomous agents which coordinate their actions to drive the robot. Our agents are fairly fine-grained; there are tens of agents all operating in parallel rather than a tightly linked set of two or three specialized sub-programs. Graphically, the mind of our robot is a schizophrenic collection of impulses that compete for control of the body. This is similar, at least in spirit, to Minsky's Society of Mind [Minsky 86]. There are many features which makes such systems attractive. The primary one is that, should an agent fail for any of a number of reasons, the whole system exhibits only a slight degradation in competence. The second advantage of multi-agent systems is that they are more easily extended than monolithic centralized controllers. For instance, if we are reasonably careful about interactions between behaviors we can incrementally augment the system's capabilities by simply adding new agents. Furthermore, since there are no central bottlenecks or shared resources to saturate, as we add agents we can also add more hardware. This lets us handle the increased demand for computing power without compromising the performance of the rest of the system.

However, from our earlier experiences with Brooks's subsumption architecture [Brooks 86], we discovered that to reap these benefits the robot's task must be broken up carefully. First, it is critical that the various agents be mutually independent. It is a violation of modularity for one agent to depend on the internal structure of another as this would preclude replacing an agent with a new improved version that was implemented differently. It would also prevent us from compiling agents down to more efficient units unless these new units left all the proper internal signals accessible. Most importantly, it would force us to stick with our original task decomposition, even if it later became unwieldy. Typically these control systems are built by successively adding new groups of behaviors on top of the existing set. Given this structure, suppose we reorganized the functions of several lower level agents. Many of the higher level agents might then have to be changed, too, because the lower level subunits they rely on would no longer be available.

The other guideline for proper partitioning is that all subtasks should require only local information to be successfully completed. This follows from the fact that, in practice, the sensory information available to the robot is usually so poor that detailed representations are nearly impossible to build. Furthermore, since the larger details of the task are obscured, algorithms based on locally perceived features of the situation are likely to be more robust in complex, crowded environments. For similar reasons, decisions should be temporally local as well. It typically does not work to take a "snapshot" of the world, devise some action sequence using this information, and then blindly follow this plan. For instance, in plotting a trajectory through space certain obstacles may not be evident at the outset, other obstacles may enter the path as the robot moves, and control errors may accumulate to the point where the robot is operating in a largely imaginary world. To have any hope of success the execution of the plan must be monitored along the way, and corrected as new sensory information becomes available. In the limit, in a highly dynamic world such a plan would have to be reformulated at each instant anyhow.

These principles, independence and locality, have important ramifications for system design. Of the two, independence is the more stringent restriction. The most obvious consequence of this choice is that there can be no complete internal world model. First, with a distributed system there would be no good place to store such a data structure. Furthermore, because all of our agents are independent and have no communication interconnections, we can not even designate one agent to hold this information and let other agents query it. This means blackboard based approaches are inappropriate. The last alternative is to have each agent build and maintain its own version of the world model. While this would be possible, it certainly does not seem practical. Also, there is the danger of skew between the copies. The whole point of a centralized world model is to allow the robot to deal with its environment in a coordinated way. Now, however, if there are multiple copies of the world model there is no guarantee that they will be identical. Thus, we lose one major advantage of the world model anyhow.

Our solution to this conundrum is to <u>use the world as its own representation</u>. We came to this conclusion by considering what world models are used for. Typically one integrates all the data from a variety of sensors to recreate internally a miniature replica of the current surroundings. Then various subroutines use this diorama to measure particular quantities and ascertain certain relationships in order to decide what action to take. Why expend all this effort to build up a model and then only use small pieces of it. Why not just measure the interesting quantities dircctly from the world? In other words, use the same action analysis routines but substitute the perceived world for the representation. No longer do we have to extract every iota of meaning; instead, each agent merely computes as much as it needs to make its control decisions (cf. [Wehner 87]).

One often cited reason for having a world model is that it allows us to combine information from different modalities, or from the same modality over time. We can still do this in our system, but we are relieved of the burden of making such mergers fine-grained and globally consistent. Sometimes, however, even this effort is not necessary. With a clever decomposition of the problem, we may only need to look at a single sensor to control some parameter of a system (such as Raibert's balancing robot [Raibert 86]). We can then combine a number of these simple systems through behavior fusion to achieve the same result as would have be achieved with sensor fusion.

The second requirement, spatio-temporal locality, also dictates a "reactive" style of programming. The robot responds solely to events in its world; it is not driven by some internal schedule. We have taken to this to an extreme and made the radical assumption that an interesting robot can be built using no persistent state. The primary benefit of this is that we never have to cold-boot our system to wipe out erroneous information. However, if we are not careful, the robot suffers from myopia and loses the overall picture of what it is supposed to be doing. If we were to relax this stricture, the robot could use a history of past events to disambiguate locally identical stimulus patterns. It could also use more global criteria to choose between actions, even if the relevant context information was not directly perceptible at the crucial moment. Yet, to confidently add state to the system requires placing a lot of trust in the robot's sensory capabilities. Since later actions and perceptions are influenced by state, we want to verify that the remembered data is based on solid readings and is not just a sensory glitch. Furthermore, since the robot may not be able to observe the same features in the future, we must have faith that it has interpreted the stimulus correctly. Unfortunately, predicting how a given sensor will react when placed in a real environment is a notoriously difficult problem.

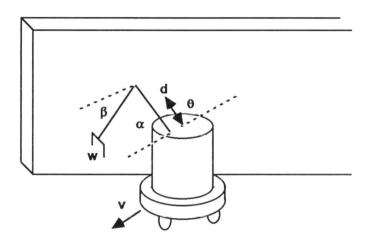

Figure 1-5. There are many sources of state external to the robot that can be used for the purposes of memory, sequencing, and communication between agents.

While we attempt to keep the amount of internal state to a minimum, the creature can still behave *as if* it had memory. As shown in figure 1-5, there are many forms of state external to the robot's control system that can be coopted for productive functions. Variables such as the arm's position, the base's speed, and the robot's orientation relative to obstacles in the environment are all potentially useful. These conditions can be used to indicate and perpetuate the current operating mode of the robot, and to signal transitions between different activities in a sequence. Although we forbid agents to talk directly to one another, often one agent will maneuver the robot into a situation which will cause the activation of another agent. In this respect, these variables also allow agents to <u>communicate through the world</u> and coordinate their actions. Many times one behavior can key off some side effect of another behavior. This obviates the need for any special internal communication channel or message drop which, in turn, helps preserve the modularity of the system. Introducing new interface primitives, such as a *retract-arm-now* signal, could require us to rewrite the lower levels of the control system to conform to this altered effector protocol.

1.4 Contributions

This report makes several contributions to our understanding of mobile robots and ways to control them. The most important accomplishments are:

Functioning Robot - A complete, operational robot was built which successfully accomplishes a difficult task in an unstructured environment.

Distributed Architecture - The structure of a practical and theoretically interesting distributed architecture was completely specified.

Limitations of Locality - A number of experiments were performed to investigate the power of spatio-temporally local control schemes.

Limitations of Arbitration - An analysis of a simple fixed priority behavior arbitration scheme was presented and its flexibility was analyzed.

Useful Algorithms - A number of novel recognition, manipulation, and navigation algorithms were developed and implemented.

In addition, there are a variety of other items which should prove of interest to other robot designers. Among these are:

Hardware Spin-offs - A large number of sensors and other hardware subsystems were designed and characterized. These include several types of infrared proximity sensors, a simple laser range finder, a low power arm with a large workspace, a network of parallel processors, and a compact pipe-lined vision system.

Examples for Analysis - Several large control systems were constructed for various aspects of the can collection task. These are concrete instances of distributed systems which can serve as models for other experimenters, and as examples for theoreticians.

1.5 Roadmap

The rest of the report is broken into three major portions. Chapter 2 sets forth our parallel control system and compares it to other existing systems. Chapter 3 describes the robot's arm and details how it finds and grasps cans. Chapter 4 is concerned with the laser light-striper and presents the recognition algorithms used and explains how the vision subsystem is interfaced to the rest of the robot. Chapter 5 examines the local navigation and strategic control components of the robot and documents their performance. Finally, Chapter 6 summarizes our empirical findings, discusses the system's limitations, and suggests avenues for future research.

Chapter 2

Architecture

The control system we have developed for our robot is modelled after the behavioral network of the snail as presented in Chapter 1. Here we first specify the exact function of each of the basic components of our system and show how they evolved from an early version of Brooks' subsumption architecture [Brooks 86]. Next, we present the multi-processor hardware used to implement the control system onboard our robot. Finally, we compare this new architecture to similar systems which have been used to control mobile robots.

2.1 What we use

A typical example of the architecture is shown in figure 2-1. It consists of a number of modules (boxes), each of which implements some small piece of behavior. These are behaviors corresponds to primitive activities such as the positive geotaxis and negative phototaxis discussed in the snail example. In this sense, modules are complete, self-contained control systems which use various types of perceptual information (P_n's) to generate appropriate commands for the creature's actuators (A_n's). Competing commands are merged using a hardwired priority scheme (circles) which is basically a series of switches. Only one module at a time can gain control of the contested actuator resource. Notice that, aside from the arbitration network, each module is independent from all the others. There are no direct channels between modules nor is there any central forum for communication. In particular, there is nothing like the globally accessible and modifiable short-term memory buffer typically found in a forward chaining production system.

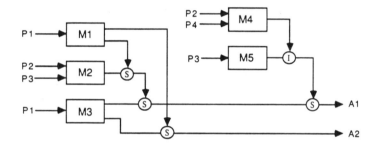

Figure 2-1. Our control system consists of a number modules (the M's), each of which implements a particular behavior. These module use the available sensor primitives (the P's) to directly generate commands for the actuator resources (the A's). The outputs of different modules are combined through a fixed arbitration network, represented here by circles.

The communication that does occur in this architecture is either from the sensors to the modules, or from the modules to the arbitration nodes. These messages travel over the "wires" shown in the diagram. Conceptually they are continuous signals that always reflect the current value of a command directive or sensor reading. For implementation reasons, however, we simulate them using discrete packets of information. When active, modules continually send packets detailing the action they desire the robot to perform. The receiver of such a stream acts on the basis of a particular packet only until the next packet arrives. A packet therefore has a finite, but very limited, temporal extent. Although each module must remember each packet for a small amount of time, theoretically the communication scheme requires no saved state. Furthermore, if packets arrive with sufficient frequency, the system is essentially continuous.

So far we have discussed the mechanics of communications but have said nothing about their content. In general, the raw sensor signals undergo various types preprocessing, either inside the module or beforehand. Typical transformations include amplification, thresholding, spatial differentiation, and noise elimination. Likewise, the basic outputs of modules are usually not actual motor currents or joint velocities. They are a shorthand for one of a few standard motion patterns which are expanded by a lookup table before being sent to an associated servo system. Again, these interpretation steps can occur either internally or as a separate post-processing stage. Our architecture is designed primarily to fill the gap between low-level sensor information and structured motor actions. In the examples given we have been careful to point out exactly what types of pre- and post-processing occur.

Now assume for a moment that we already have a number of modules and want to specify how they should interact. Two special constructs are provided for this purpose. First, a module can inhibit the output of another module (circle with an "I") and so prevent it from generating any outputs. Second, the output of one module can suppress the output of another (circle with an "S"). In the case of suppression, the output from the dominant module overrides the output of the inferior module. Not only are the inferior module's commands blocked, but the dominant module actually substitutes its own commands in place of original commands. This is a particularly powerful construct because it lets more competent behaviors take over from general-purpose behaviors when the circumstances warrant it.

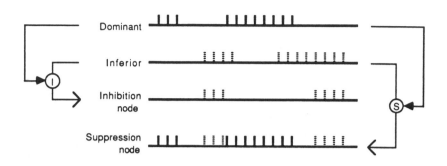

Figure 2-2. These are the timing diagrams for inhibition and suppression nodes operating on streams of packets. Notice that the suppressor node substitutes the stream from the dominant module (the dark colored blips) in place of the inferior module's output.

Figure 2-2 shows the effects of suppressor and inhibiter nodes. Since each output is really a stream of packets, they are represented here as a series of spikes over time. When the dominant module sends packets (top line) into an inhibition node, the inferior module's packets (second line) no longer make it to the output (third line). The dominant module also blocks the inferior module's packets in a suppressor node. However, in addition the dominant node injects its own packets (fourth line) onto the wire normally used by the inferior module. If we were dealing with true signals, the dominant module would lose control of a node as soon as it stopped generating an output. Yet with our serial encoding scheme there is short gap between packets in a stream. To compensate, we require each node to remain active for a small amount of time after the last triggering input. If a another packet does not arrive before this interval expires, the node switches back and the inferior module's commands pass through unaltered. Thus, conceptually, arbitration nodes contain no state.

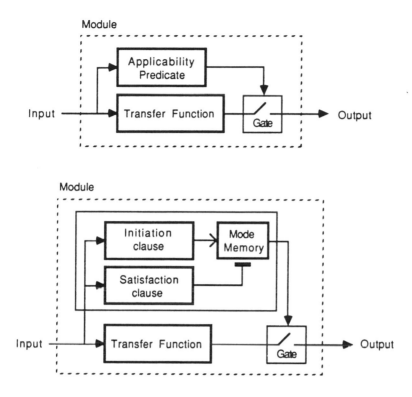

Figure 2-3. Inside a module the transfer function describes what action the module will take while the applicability predicate decides when to gate these commands to the output. Sometimes the the applicability predicate is given a small amount of state as well (bottom).

Modules themselves can be considered roughly equivalent to production rules or situation-action responses. Their internal workings, shown in top half of figure 2-3, consist of two major parts which together give rise to each module's unique behavior. The first part, the transfer function, defines what sort of action to take based on the sensory input to the module. For instance, recall the control system for the snail discussed in Chapter 1. The mechanism that oriented the creature toward bright areas is an example of a transfer function. The other part of a module, its applicability predicate, determines when the transfer function should generate commands. Again from the snail example, recall that the animal only sought light when it was upside down. In this case the applicability predicate would be a circuit that detected inversion of the creature. Only under these special conditions is the result of the transfer function gated to the module's output.

In many cases the applicability predicate is used as a goal statement and the transfer function alters the world so that this predicate becomes false. For instance, if there is an object directly ahead of the robot, a module which turns the robot to the side becomes active. Eventually, this action causes the object to no longer be in front of the robot and thus the applicability predicate for this module becomes false. The robot has achieved the "goal" of escaping from the situation which triggered this turning behavior. However, not all the behaviors in our robot take this form. Sometimes the applicability predicate is related to a desirable goal state and the transfer function acts to maintain this state. Other times the transfer function's action does not affect the applicability predicate at all. The applicability predicate serves to simply switch in some new behavior in a certain situation which may or may not persist as a result of this addition.

Sometimes, however, the robot needs to respond to events as well as situations. The distinction we draw between these two classes is that situations are extended intervals of time whereas events are isolated point-like occurrences. In the case where the module is to be triggered by an event we need to stretch the duration of the applicability predicate in order to allow the module to have time to influence the robot. For this types of behaviors we split the applicability predicate into three parts as shown in the lower half of figure 2-3. The initiation predicate detects the triggering event and sets the mode memory to true.

The satisfaction predicate performs the opposite function and resets the mode memory when it detects that the goal has been achieved. We store this single mode bit in a special type of memory latch called a retriggerable monostable. As shown in the timing diagram, figure 2-4, when its input goes high this unit remembers one bit of information. However, if its input has been low for a preset length of time, it automatically resets itself. This prevents possibly outdated information from lingering on and exerting undue influence on the operation of the robot. Using this construct, it is as if every module had a built-in watchdog timer that "booted" it at regular intervals. Of course, if the proper termination event occurs within this interval, the satisfaction clause can directly clear the mode bit before its timer expires.

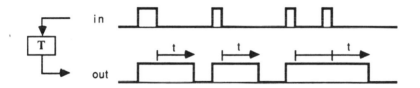

Figure 2-4. The timing diagram for a retriggerable monostable is shown here. This construct can be considered a piece of state which automatically resets itself after a while.

The gating function used inside modules is actually very similar to an inhibition node. In some sense then, the transfer function and applicability predicate can be viewed as two separate modules. The transfer function would be a module that continuously generated commands and the applicability predicate would be another module that inhibited the first in most cases. When the appropriate situation occurs, however, the applicability module would cease producing outputs and would thus release the transfer module to act. Yet instead of pursuing this approach, we decided to lump the two pieces together for a number of reasons. First, the output of the two modules are of different abstract data types. For inhibition the applicability predicate only needs to generate a single binary output, whereas the bandwidth of the transfer function is much higher. Second, this is a recurrent pattern of interaction that is present in many places within a typical control system. It is also just about the only place inhibition is actually used. For our robot at least, the entire arbitration network consists exclusively of suppressor nodes. Finally, we have found from experience that applicability predicates usually contain most of the interesting processing. Transfer functions are often very simple or even invariant such as the function "go forward".

2.2 The subsumption architecture

The distributed control system we use bears a high degree of similarity to Rod Brooks' subsumption architecture [Brooks 86]. This correspondence is natural since our approach evolved from subsumption architecture, and represents a refinement on the same basic ideas. Brooks himself has now adopted many of the ideas developed in this report [Brooks 89].

The major structural difference between the two architectures centers around Brooks' particular method for decomposing his control systems into layers. Both approaches build up controllers by incrementally adding new levels to an existing system. However, Brooks envisions this as a uniform process over the entire system in which all aspects of control are improved simultaneously. We regard our own architecture more as a "soup" of modules than as a stratified heap. This difference might be summarized by saying that Brooks' layers define a total order on the behaviors of a robot, whereas ours only define a tree-like partial order. For instance, the controllers for different actuator resources are almost always unrelated. Even within groups devoted to the same resource there are often several disjoint branches. Also, contrary to Brooks' original proposition, we do not require the dominance of various layers to follow their evolutionary sequence. A new layer might provide a weak general purpose solution that should only be used when the more specialized lower layers do not know what to do. In fact, we even allow the various portions of a layer to have *different* priorities relative to whatever existing layers they interact with.

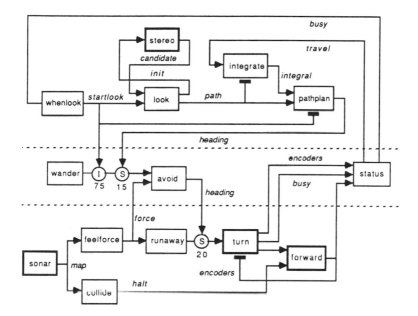

Figure 2-5. This is the control system for Brooks' first robot. Higher levels spy on wires contained in the lower levels and sometimes inject their own signals.

Moreover, in our system when dependencies between different layers occur, they involve the function performed by a layer rather than its internal structure. By contrast, one of the prime features of the subsumption architecture is that upper levels can "spy" on the connections of lower levels and "inject" alternative signals onto these paths. Figure 2-5 shows one of

Brooks's control systems with the levels separated by dashed lines. As can be seen, there are several connections that cross these level boundaries. This certainly violates our design princi- ple which calls for independence of modules. As mentioned before, the major problem with this style is that the designer must view the system holistically (or just be lucky) and choose the correct decomposition for the lower levels at the start. Otherwise the proper signals and injection points for higher levels may not be available.

Another divergence between our approach and that of Brooks concerns the semantics of suppression and inhibition. In Brooks' early system each suppressor and inhibiter node has a time constant associated with it [Brooks 86]. When a higher level module commandeers such a node, the module retains control for a prespecified length of time. This is a result of the fact that each of Brooks' message packets is meant to provide control of the robot for some non-trivial interval of time. To drive the robot a certain distance one would send a single "go forward" message and then follow it, some time later, with a "stop" message. Thus, instead of being a pure reflex response, each packet really contains a short piece of a plan. Unfortu- nately, suppressor nodes do not work well with this communi- cations protocol. As shown in figure 2-6a, if some module (M1) grabs a suppressor node it blocks any other commands until its task is complete. This can prove catastrophic if the losing contender for this node (M2) generates a packet of its own during this interval. For instance, M2 might generate a "stop" command **only** when an obstacle first enters the robot's projected path. Looking at the output of the suppressor node (R) we see that this command would be lost and never reach the effectors. Figure 2-6b shows the same situation using our new signal model of communication. The lock-out problem is eliminated because the inferior module is still generating outputs when the dominant module relinquishes control.

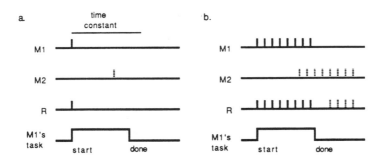

Figure 2-6. a. In Brooks' packet model, M1 takes control of the suppressor node for a preset length of time. A command generated by M2 during this period will never reach the output of the suppressor node (R). b. With our signal model, M2 is still generating a stream of packets when M1's task completes. Thus its request eventually gets serviced.

In general Brooks relies much more on state than we do. In fact, his modules are explicitly made of augmented finite state machines (AFSM's) which are basically very small sequential programs. Modules are also allowed to include "instance" variables and time constants. Thus there are three classes of state: sequencing steps, variables, and timers. The variables are undeniably state because they persist indefinitely. The timers, on the other hand, are similar to the monostables provided by our system and thus are "safe" state. Somewhat surprisingly, the way Brooks uses the AFSM formalism does not actually give rise to very much *persistent* state. The reason modules sequence between entries is that, by Brooks' language definition, only one operation can be accomplished per entry. Shown below is the code for the RUNAWAY module's state machine. The first entry, the "event dispatch", merely waits for a packet containing the digested map information to arrive. The next entry, the "conditional dispatch", tests the resultant force vector to see if it is big enough to prompt action. If so, the last entry, an "output", generates a command which causes the robot to run directly away from the disturbance. Notice that all these states are ephemeral; they are only used to synchronize the flow of data.

```
(defmodule runaway
   :inputs  (force)
   :outputs (command)
   :states
     ((nil    (event-dispatch force decide))
      (decide (conditional-dispatch
               (significant-force-p force)
               runaway
               nil))
      (runaway
        (output command (follow-force force))
        nil)))
```

This module also maps fairly directly to our new language. SIGNIFICANT-FORCE-P corresponds to the module's applicability predicate while FOLLOW-FORCE is its transfer function. The remaining step in the original state machine, the event dispatch, would not be needed since we have switched to a signal model of communication. The converse mapping from our architecture to subsumption is also straightforward. In general, one of our modules can be represented as a pair of state machines in Brooks' formalism. As shown below, the module waits for a sensory packet to arrive then runs the applicability predicate on it. If the relevant conditions are not satisfied the module jumps back to the first state and waits for another packet. It stays in this first loop until a packet arrives which triggers the applicability predicate. It then jumps into the second loop and emits a signal which resets the mode memory. Next it runs the transfer function on the sensory data to generate a command output for the actuators. Finally, it jumps back to the first state in this loop and waits for more inputs. The module will stay in this new loop indefinitely; thus the state machine has effectively recorded the fact that the module was once activated.

```
(defmodule basic
  :inputs  (sensors)
  :outputs (mode actuators)
  :states
    ((nil (event-dispatch sensors start))
     (start
       (conditional-dispatch
         (APPLICABILITY-PREDICATE  sensors)
         renew
         nil))
     (wait (event-dispatch sensors check))
     (check
       (conditional-dispatch
         (APPLICABILITY-PREDICATE  sensors)
         renew
         active))
     (renew
       (output mode $true$)
       active)
     (active
       (output actuators
               (TRANSFER-FUNCTION  sensors))
       wait)))
```

Notice that every time through this second loop, the module runs the transfer function and generates an output. However, it generates a mode memory output only if the applicability predicate is still satisfied. We send this output to a second module which implements the mode memory monostable. When a trigger signals arrives, this module jumps to the second event dispatch state and starts up an internal timer. If another trigger impulse arrives during the specified interval, the "sleep" construct is reinitialized. Thus, as long as the applicability predicate in the first module is true (and for a short while afterward), the second module will remain in this event dispatch state. However, when the timer finishes counting the module will emit a single output pulse before returning to its initial state. This pulse goes to a special reset input on the first module and forces the internal state machine to go to the first state. Thus, when the second module times out, the first module switches out of its output producing loop and goes back to its original testing loop.

```
(defmodule watchdog
  :inputs  (trigger)
  :outputs (reset)
  :states
    ((nil (event-dispatch trigger stretch))
     (stretch
       (event-dispatch
         trigger stretch
         (sleep MONOSTABLE-TIME)  switch))
     (switch
       (output reset $true$)
       nil)))
```

2.3 The multiprocessor implementation

On an earlier robot [Brooks and Connell 86] we tried using telemetry and off-board computation, but found this to be highly unreliable. For this reason we decided to go with all on-board computation on this robot, Herbert. Instead of using a production system or discrete logic as mentioned before, we chose to implement all his behaviors on a set of loosely-coupled 8 bit microprocessors. These were designed to run both Brooks' original version of the subsumption architecture and the new version presented here. Physically, these processors boards are 3 inches by 4 inches and, except for power, are completely self-contained. The core of each board is a Hitachi HD63P01M1 microprocessor; essentially a CMOS version of the Motorola 6800 with 128 bytes of RAM and a piggyback socket for an 8K EPROM. The processor configured to run with external memory and we have left a provision to allow an additional 2K of static RAM. In practice, however, this extra memory has proved unnecessary so we have removed it. All the components on the board are CMOS to keep power consumption low (about 150 mw). This is a critical property for a mobile robot which runs off batteries.

Each of the processor boards has a small amount of space left over. On standard boards this is devoted to a hardware implementation of a suppressor node. The time constant associated with this node can be adjusted using a potentiometer located at the corner of the board. Since we prefer the signal model of communication, we simply set this control to slightly longer than the standard interpacket time and leave it there. On other boards, called "mutants", the extra space is instead used to implement an 8 bit bi-directional parallel port. This port is mapped to a group of 4 memory addresses and thereby allows the processor to communicate with a number of peripheral devices.

There are also several features which are important for Brooks' modules but are not used in our system. One is the reset line. If any message arrives on this input an interrupt is generated and the processor reinitializes all its internal finite state machines. Since our modules have virtually no state, this input goes unused. Another obsolete feature is a built-in inhibition input which is checked by the processor every time it wants to generate an output. When active, it forbids the module from sending packets on any of the outputs. Thanks to special hardware, this line can remain active for a short time after the last triggering input arrived. The actual period can range up to a minute and, like the time constant of the suppressor node, is controlled by a potentiometer at the top of the board. However, since our arbiter uses suppression nodes exclusively, the inhibition line and its associated hardware are also superfluous.

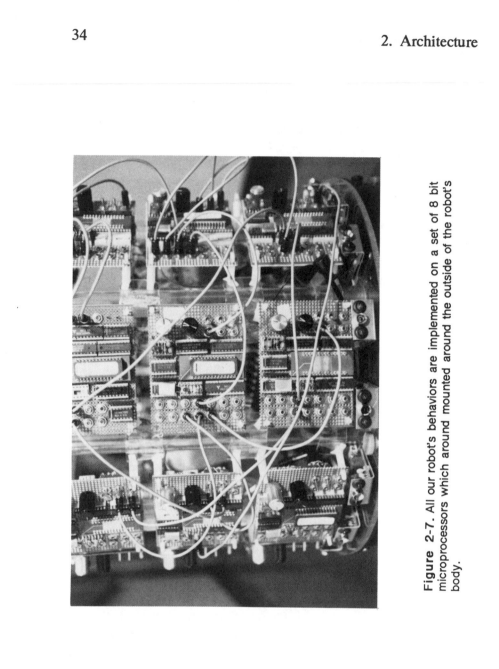

Figure 2-7. All our robot's behaviors are implemented on a set of 8 bit microprocessors which around mounted around the outside of the robot's body.

Processors communicate with other processors over special two-wire serial connections. These connections carry 24 bit packets at an effective rate of approximately 300 baud. Each board provides 3 simultaneous serial inputs and 3 serial outputs. The outputs go to single subminiature phone jacks while the inputs come from paired jacks to allow for the daisy-chaining of signals. Although the 6301 microcontroller does have a built-in serial port, it has only one. Therefore, we synthesized the six necessary channels by polling several parallel input ports in software. The program that does this consumes 80% of the processing power available on the board. However, if each instruction takes 3 microseconds and there are 10 packets a second, this still lets the processor perform over 6000 operations per packet.

The communications protocol we use allows the clocks of the sender and receiver to differ by as much as 20% and to have an arbitrary phase relative to each other. The processor starts transmission by momentarily pulling the associated strobe line low. This signals the receiving processor that a new packet its about to arrive and toggles the state of any intervening suppressor nodes. It also explicitly tells the receiver to discard any partial packet that may have been in progress. This is important because a suppressor node can switch from the inferior module to the dominant one right in the middle of a packets. Such a truncated packet is of no use and should be ignored. After the synchronization pulse, the actual packet is sent on the data line using a form of pulse length modulation. Ones are converted into long pulses while zeroes are represented by shorter blips. Thus a message with many ones takes longer to transmit than a packet containing all zeroes.

Each processor is wired, by hand, only to the other processors it needs to talk to. As can be seen, this leads to a morass of small gray cables that cover the surface of the robot. However, there is no central bus, backplane, or blackboard for general purpose communications - only this distributed patch panel. This is important because it means there are no information bottlenecks or shared resources to saturate. As a consequence, the hardware can be extended *indefinitely* without degrading the system's performance. If we want to implement more behaviors we simply add more processors.

The actual processor boards are mounted on the surface of the robot to allow easy access. There is space around the periphery of the robot for 24 processor boards, but some of these slots are filled by other types of hardware (such as mutants). Originally we had intended to map directly from modules to processors. However, since we have 41 behaviors and only 13 positions actually available, we must instead package 3 or 4 behaviors per processor. In general, we accomplish this by grouping together related behaviors into "levels of competence" and assigning one of these per processor. Since each board has a manual shutdown switch which disables the processor, this allows us to see what affect each level has on the overall performance of the robot. Unfortunately, in our new style of subsumption we typically need a suppressor node for every module. Since we don't have enough hardware for this, we internally simulate all the suppression interactions between modules within a small task-oriented group. The output of this arbitration then goes to the hardware suppression node to be combined with commands from other clumps of behaviors.

2.4 Related architectures

The architecture presented here also bears similarities to many of the multi-agent control systems proposed by other researchers in mobile robotics. The closest of these is probably the Denning mobile robot system [Kadonoff et al. 86]. This robot has a number of different navigational skills that are switched on or off depending on the environmental conditions. For instance, normally the robot homes in and follows a modulated infrared beacon. This can be considered the behavior's transfer function. Yet the robot only follows a beacon if the beacon is relatively far away. This criterion, therefore, is the behavior's applicability predicate. As in our system, both of these portions are stateless.

The Denning robot also concurrently runs a number of other navigational processes. For instance, if the robot ever deviates significantly from its nominal path, a dead-reckoning scheme kicks in to correct the drift. As with the previous behavior, this strategy is clearly composed of an applicability predicate and a transfer function. More importantly, however, when the dead-reckoning system is active it completely overrides the beacon follower. Similarly, a third, sonar-based system which takes care of local obstacle avoidance. If an obstruction appears within a certain distance of the robot, this system slows the vehicle and turns it away from the stimulus. As before, when activated this process simply grabs control of the robot from whichever other behavior was previously driving the wheels. Denning's arbitration system merely switches between processes in a preordained hierarchy; it never combines commands. In this respect it is identical to our networks of suppressor nodes.

Researchers at SRI have also proposed a multi-agent control scheme for mobile robots [Kaebling 87]. Their system differs from ours in that it is split into a perception component and an action component. In the perception half there are a variety of strategies running in parallel which analyze the outputs of robot's sensors and cooperatively detect the presence of certain high-level perceptual primitives. The action half, like our system, is broken down into a number of totally independent components which generate effectors commands when the appropriate conditions exist. However, these components do no sensor processing themselves, but instead have access to all the details of the world model generated by the first half. The commands produced are then passed through a series of "mediators" which combine them into an appropriate overall drive signal for the robot. In the example given, Kaebling uses a fixed priority scheme like ours. However, in general, she allows arbitrarily complex arbiters to be used. Likewise, there are no restrictions placed on the nature of the computations performed by either the perception or action modules. In fact, they mention embedding planners in various subsystems.

At Hughes, they have also investigated dividing the control system into separate perception and planning units [Payton 86; Wong and Payton 87]. However, instead of having a number of perceptual subroutines which cooperate to build a total world model, Payton and his colleagues have what they refer to as "virtual sensors". These are a series of partial world models which are often aimed at detecting very specialized environmental features. For instance, one source of information used in their control system is an obstacle recognizer which signals when the robot is about to hit something. This information does not necessarily come from one sensor, nor is it the product of a single isolated perception routine. Instead, it may be the merged result of several different processes operating on a variety of sensory modalities such as proximity detectors, sonar, laser scanners, whiskers, bump switches, or stall sensors. A virtual sensor is defined by what it registers instead of how it actually accomplishes this.

The outputs of these virtual sensors serve as the inputs for the action component. Again, like our system, this is composed of a number of independent reflexive behaviors which interact (usually) through a fixed priority arbitration scheme. However, Payton also grafts a meta-level onto this system to allow for more intelligent, although slower, control. He does this by creating groups of behaviors called "activation sets". These are composed of several behaviors along with special parameter settings for certain reflexes and an arbitration order for selecting between them. Individual behaviors can belong to more than one set, but only those behaviors that are members of a currently valid set are allowed to run. The activation sets are then manipulated by a local planning module which decides when to switch on and off the various groups of behaviors. This is similar to the concept of "K-lines" as proposed by Minsky [Minsky 80; Minsky 86].

The idea of suggesting processes for the robot to use, as opposed to actions for it to execute, is also advocated by researchers at the University of Minnesota [Anderson and Donath 88a; Anderson and Donath 88b]. Using simulations they have experimented with activating only certain subsets of the available reflexes to achieve robust robot navigation. Unlike Payton's system, however, their behaviors do not have parameters that can be adjusted and the relative strengths are the same irrespective of the particular group selected. Furthermore, instead of resorting to a classical planner, behavior sets are selected using pre-wired propositions. These propositions are similar to our applicability predicates but, because they are tied to groups, typically controlling more than one transfer function. The researchers present a simple example in which the robot is attracted to a certain global position while being repelled by local obstacles. Although the robot eventually reaches its destination, it tends to jitter around the selected point when it gets close. To cure this, much as in a simple feedback system, they simply switch off the local attraction behavior at the meta-level when the robot gets sufficiently close to its goal.

The underlying reflex system in this architecture is also interesting because it adheres to some of the same principles we propose. In particular, Anderson and Donath take care to keep each of their behaviors independent of all others, and explicitly forbid them to use any form of state. The outputs of various modules, however, are not arbitrated by a priority based scheme. Instead, as in Brooks' early work [Brooks and Connell

86], the commands are combined using a vector summation technique. Each competing direction command is interpreted is as a force vector in space. The robot then computes the resultant of all such vectors to pick a direction of travel. The University of Minnesota architecture is also interesting because it contains a unique repertoire of behaviors. Included are such primitives as "forward attraction" (go in the direction the robot is facing), "narrow open space attraction" (find door-like gaps), and "location directed open space attraction" (head toward the open space most nearly aligned with the goal). These allow the robot to pursue a wider variety of trajectories than are possible with other systems, such as Arkin's [Arkin 87].

Chapter 3

Manipulation

This chapter describes the control system for the onboard arm. We rely on a series of reflexes to guide our robot. The set of behaviors developed here allows the robot to acquire and retrieve soda cans by groping around on supporting surfaces. The robot starts by raising its hand to the top of the workspace, extending the arm slightly, then bringing the hand straight down to find the table. If the fingers touch a surface, they recoil and the hand starts hopping along in search of a can. When the hand gets near, local proximity sensors align the can and provide guidance for the terminal grasping motions. Finally, the robot lifts the can straight up off the table and brings it back to a parking location next to the body. As long as the target is upright and within the workspace (3' high by 1' long by 3" wide) the arm will eventually find and retrieve it.

This control system demonstrates two important principles. First, the robot's trajectory toward the can is guided by the environment itself, rather than by some plan developed from an internal model. The robot contains a variety of different behaviors which respond to particular local aspects of the

Figure 3-1. The arm mounted on the top of our mobile robot is used to collect soda cans. Despite working in a cluttered, changing environment it is able to perform this task using simple sensors and coarse manipulator control.

situation. As the hand moves, the changing properties of the environment encountered are used to selectively activate appropriate behaviors. This arrangement lets the robot negotiate even highly cluttered areas without assembling a detailed world model. This is an advantage because comprehensive models are difficult to construct and typically make no provisions for environments that change dynamically. Systems such as [Ikeuchi et al. 86] stare at the scene for a long time then compute a ballistic trajectory for the arm. If the world is altered after the initial survey period, the robot may blunder into unexpected situations that prevent it from executing the specified plan. The multi-agent approach we use, on the other hand, is able to "improvise" since its reactive system only makes short range plans to begin with.

The other important principle is that even a system composed of independent local agents can exhibit globally directed behavior. Since our system primarily uses spatially and temporally local information, it occasionally gets caught in loops or local minima. Yet, to overcome this we do not need a central supervisor that notices certain actions are being repeated or that the improper minimum has been attained. Instead, we add another agent which monitors some sensory variable indicative of the overall progress of the robot. When this agent notes that the current set of behaviors is failing, it switches on a new set of behaviors to help the robot escape from the stagnated situation. This approach has the advantage that it allows us to build working systems without incorporating large amounts of self-knowledge into them. Requiring information about how a certain task is accomplished violates the modularity of the system and locks us into a particular implementation.

As an example, in our arm controller there is an agent that monitors the robot's groping activity by continually checking whether the hand is still moving across the workspace. If for some reason the hand stops or oscillates between nearby positions, control is transferred from the extension mode into the retraction mode. Yet this agent that does this not need to know why the hand stopped. The hand might have recently acquired or deposited a can, reached the edge of the workspace, or simply gotten stuck. Similarly, it does not even need to know which groping behaviors were active at the time. By using sensor derived events as signals we retain flexibility in the actual construction of the system without sacrificing our capability to provide global direction.

3.1 Hardware

Before examining the controller it is first necessary to understand the capabilities and limitations of the manipulator itself. We initially considered using a commercial arm but rejected this idea for two reasons. Since we wanted the robot to work at a variety of heights, we needed an arm with a large workspace. Unfortunately, such arms are so heavy that they would tip the robot over when extended. Furthermore, they are so bulky that they would not fit well on the mobile platform we had chosen. Finally, they consumed a substantial amount of power, both for the motors and for the required controller boards. None of them was suitable for running with batteries. There were, however, a variety of smaller arms that would fit on the robot and come within its power budget. Yet these arms had very small workspaces (on the order of a cubic foot) and small payload capacities (typically several ounces). Thus, we decided to build a custom manipulator for our robot. Our final design has a large workspace, weighs only 10 pounds, and can lift over a pound at full extension (albeit slowly). Both the arm and its controller fit easily onboard our robot base and consume only 18 watts on average (32 watts when heavily loaded).

The actual arm used in our studies is shown in figure 3-1. It is formed from two planar parallelogram linkages and resembles a desk lamp. These linkages serve to keep the fingers at a constant orientation relative to the mounting plate. Figure 3-2 shows all the places that the finger tips can reach. As can be seen, the robot is able to grasp objects both on the floor and at table top height. Our controller deals primarily with a rectangle carved out of the center of the full workspace. This "logical" workspace is a vertically oriented plane 12.5" wide by 39" high which is centered 20" beyond the edge of the robot. To achieve lateral displacement of the arm the whole robot base must be rotated.

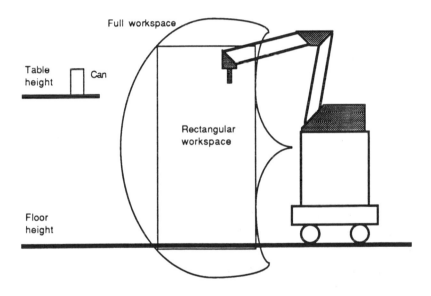

Figure 3-2. The arm is a planar two degree of freedom design with twin parallelogram linkages. The hand can reach cans both on the floor and at table height.

At the end of the arm is a vertically oriented parallel jaw gripper. This consists of a pair of 3" long fingers, each of which is surfaced with two rubber tubes. The tubes give the robot a soft, compliant grip and also provide some passive centering of the can. The fingers themselves open to a maximum width of 1.25". Since soda cans slide easily across surfaces, this wide gap lets the robot tolerate a lateral misalignment of the can of plus or minus 3/4". However, the hand is not as well matched to the height of cans. Because the fingers are short, the hand must be raised off the supporting surface in order to grab a can.

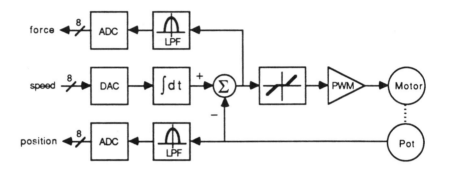

Figure 3-3. Each motor is controlled by an analog position servo whose setpoint is derived by integrating a velocity command. The position and error signals of each servo can be accessed by higher levels of control.

The shoulder, elbow, and fingers motors are each driven by identical analog position servos. The actual servo board is bolted to the side of the shoulder gear box. The setpoint comes from an analog integrator fed by either a joystick (useful for debugging) or an 8 bit digital to analog converter. The integration time constants of the shoulder and elbow servos are adjusted to give comparable joint velocities for the same command byte. The same effect could be achieved by feeding a standard position servo with a sequence of successive points on the path. However, the hardware integrator lets us lower the update rate and, for a given level of quantization, provides smoother control of the arm at slow speeds. Our system is not a true velocity controller since we do not use tachometer signals nor do we differentiate the position signal. Yet we choose to use velocity commands rather than positional ones because, with our reflex-based control system, most movements are made relative to the current location of the hand. That is, we are more interested in controlling the trajectory of the arm than in sending it to some global position.

The servo board communicates over a parallel port connection with a modified subsumption architecture processor board called a "mutant". The mutant provides an interface to the arm using standard subsumption serial lines. Its primary function is to allow cartesian control of the arm by running the appropriate inverse kinematic calculations on the received velocity vector. For various reasons the linear speed of the arm is not directly controllable and varies somewhat across the workspace. For our purposes, however, all that is important is knowing that the hand moves approximately 1.3" per second. In addition to its control function, the mutant also reports various system parameters useful to the subsumption processors. For instance, the distance between the fingers and the force applied is continuously transmitted on one of the mutant's output channels. Similarly, another output reports the x-y position as computed from the joint angles. All outputs have a 14.3 Hz update rate.

3.2 Sensors

In addition to position and force information, the subsumption architecture control system has access to a number of sensors located on the hand itself. As shown in figure 3-4, there are a variety of sensors all of which are updated at a rate of 9.8 Hz. For instance, there is an infrared beam at the back edge of the fingers roughly 1" up from the tips. This operates like the light beam in an elevator's door: the robot can detect when it is broken by an object. There are also two switches on each fingertip which signal contact with a surface. These are 3/8" in diameter and take about 5 ounces of pressure to activate.

Although the responses of all four switches are available to the control system, we typically OR the bits together to give a single control variable. The hand also has a microswitch at the back of the wrist which acts as a primitive force sensor. Since the whole finger assembly is pivoted at its front edge, the wrist switch serves to detect a torque around this axis. This sensor is triggered if the fingers are rocked backward with a torque in excess of 3 ounce-inches. The sensor will also be triggered if a significant upward force is applied to the fingertips. Since the fingertips can not reach the table when the hand is holding a can, the wrist switch is sometimes the only source of tactile information.

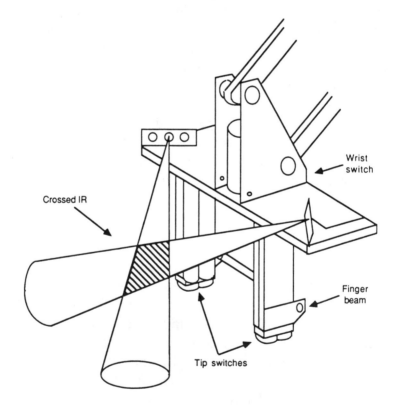

Figure 3-4. The hand has a variety of local sensors.

The most useful source of local information is a pair of crossed infrared beams located at the leading edge of the hand. The actual sensor heads are angled 45 degrees inward and angled 30 degrees down. In the basic proximity mode they operate by emitting near infrared radiation in a tight cone and then looking for how much is reflected back. This quantity is then thresholded to determine if there is something in the beam. Under favorable conditions the sensors on the hand can detect white paper at over a foot. However, due to high levels of electrical and motor noise we have tuned them down to a range of 4 inches. Figure 3-5 shows the sensing regions for different elevations of the hand above the table. This mapping was obtained by first propping the hand up the specified distance, and then moving a real soda can through each detector's field of view. We found the extent of the detection area, and then deconvolved its boundaries with a circle the same size as the can to give the corresponding positions for the center of the can. Figure 3-5 plots this along with the 1.5" wide grasp zone of the fingers.

The result was not quite what we had expected because the top of the can is highly specular and the whole curved outer surface contributes to the reflected signal. We had hoped to use the crossed structure of the sensors to provide lateralization cues for aligning the object with the gripper. Unfortunately, the noise problems severely truncated the sensor's range which in turn shortened the far field sensitivity pattern. Instead of having a large funnel-shaped extension, the side discrimination fields are now stubby and even cross over the centerline. Fortunately, this phenomenon depends on the height of the hand so the control system is able to compensate for it by keeping the hand close to the ground. The problem could be eliminated altogether by reducing the baseline between the two sensors. For instance, if both were located near the center of the hand instead of at the edges, each sensor's range would be predominantly on one side but there still would be a useful intersection region.

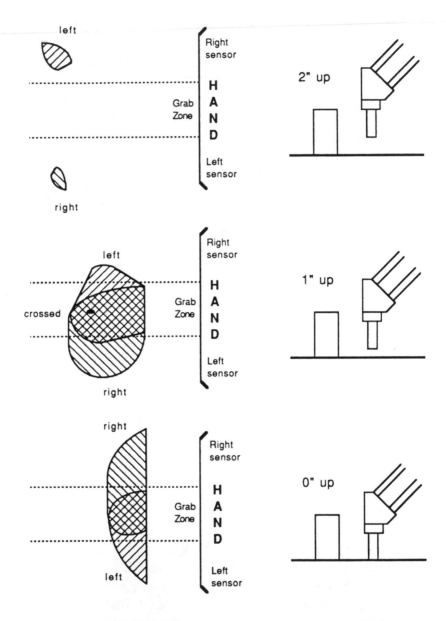

Figure 3-5. The IR sensors will detect a can if its center is in one of the shaded regions. To fit between the fingers, the can's center must lie in the marked grab zone.

3.3 Controlling the hand

The robot's hand is controlled cooperatively by a set of 6 independent behaviors. Figure 3-6 shows these behaviors and their interconnections. Each box in the diagram is a "module" and implements a specific behavior. All behaviors interact through "suppressor" nodes (circles "S" inside). As mentioned earlier, the semantics of these nodes is that a signal coming into the top can override the signal passing through. Note that we have broken the total collection down into two separate groups - **Cradle** and **Grip** - which we refer to as "levels of competence". Here, the **Grip** level allows the robot to grasp the easiest class of cans - those already between its fingers. The **Cradle** level then monitors the force applied to prevent damage to either the can or the robot. The robot can perform useful actions with just the lower level implemented, but its performance gets progressively better as more and more levels are added. This effect is more evident for the arm controller which is broken down into many more levels.

The hand controller is very simple and just serves to illustrate how an actual system can be constructed from a number of independent agents. For instance, the **Grip** level tells the robot how and when to grasp things. The most basic behavior in this level, CLAW, instructs the fingers to stay wherever they happen to be. This default behavior is modified by the GRAB module which suppresses CLAW's output (circle with an "S" inside). This module has a very simple "model" of a can: anything that fits between the robot's fingers counts. The robot could not care less whether the object is a soda can or something else; it just knows to close its fingers in this situation. The GRAB module detects such suitable objects by monitoring the light beam at the back of the fingers. If no object is sensed, GRAB instead directs the hand to open in preparation for the next object.

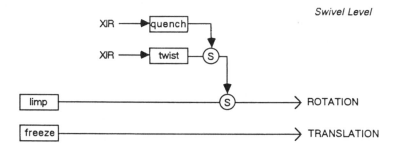

Figure 3-6. The control system for the hand is divided into two levels of competence.

The GRAB behavior is in turn modified by DEPOSIT which forces the hand to open if either the wrist sensor or one of the tip switches is activated. As shown in the diagram this module's output takes precedent over both CLAW and GRAB. This new behavior forms the basis of the robot's "plan" for putting things down: it moves the hand down until a contact between the grasped can and the table causes sufficient force to activate the wrist switch. DEPOSIT is also useful in its own right as a general protective behavior. For instance, the wrist is also activated when the bottom of the can catches on a protruding obstacle, or when the robot has grabbed a person's hand and they attempt to shake it loose. In these cases DEPOSIT, as before, causes the can to be released.

The situation is not quite this clear cut, however, and modifications to the DEPOSIT behavior are required. In particular, as soon as the fingers ease up their pressure on the object, the wrist is free to rotate and relieve the built up stress. Yet at this point the finger beam still senses the object prompting GRAB to immediately regrasp it. Therefore we have included a monostable within DEPOSIT which cause the fingers to *continue* to open for a short while after the tactile stimulus vanishes. By prolonging the opening phase, we make sure that they robot has completely released the object. We also give other behaviors time in which to act (e.g. the UNCRASH module to be described later).

The next level of control, **Cradle**, shows how a primitive force feedback servo loop can be built using our architecture. Since the gripper can exert nearly 10 pounds of force, there is a distinct danger of crushing the can unless we include a regulator such as this. However, because we do not have any component equivalent to an adder, the resulting structure of our controller is somewhat baroque. This is an example of a job for which a subsumption-style architecture is poorly suited - a traditional servo system would have worked much better. Nevertheless, this control layer is interesting because it illustrates how a judicious assignment of semantics to signals can mitigate some of the shortcomings of a fixed priority arbitration scheme.

Inside the **Cradle** level there are 3 modules which all monitor the error in the finger servo-loop, but react in different ways. While the motor servo system provides 8 bits of force data, only 2 or 3 of these bits are above the noise level. Therefore, we use only a coarse quantization of force as shown in figure 3-7. The first behavior, EXCESS, takes control when the force reaches a dangerously high level and drives the fingers in a direction which reduce the force. To bring the force into the acceptable range, this module actual continues to drive the fingers even after the initial threshold is passed (again, using a monostable). However, this module alone is not sufficient. After a brief reduction, the grasp force would normally start to increase until EXCESS was again triggered. To prevent the fingers from oscillating and applying a pulsating force, we add the CRUSH and SPREAD modules. If the fingers have been exerting a noticeable force for a while, the relevant module switches on and holds the finger force within the permitted range. The only difference between the two is that CRUSH moderates excessive grasping pressure, whereas SPREAD regulates the opening force. A monostable is used to help determine the length of time that the force has been high. While this is obviously a form of state, nothing is permanently recorded for later use. In fact, these modules are implemented so that they are triggered when the memory of an acceptable level of force has faded away.

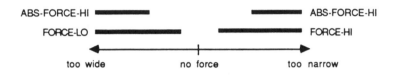

Figure 3-7. When the grasp force is in either portion of the ABS-FORCE-HI region, the fingers are prevented from moving any further. The FORCE-LO and FORCE-HI regions are used to determine the action necessary to relieve the excessive stress using a "bang-bang" control scheme.

The proper functioning of the two modules relies on being able to generate a unidirectional prohibitory effect. If we simply stopped the hand when the force was in the desired region, once CRUSH or SPREAD was triggered it would remain active indefinitely. If we were then to use normal suppression the activation of would of these modules would effectively lock the fingers in their current positions forever. We overcome this limitation by using a special two line command encoding and then *partially* suppressing the existing finger speed signal. Notice that in the subsumption diagram (figure 3-6) we show the hand control bundle splitting into its component wires at one point. The CRUSH and SPREAD modules are then free to inject signals on just one of these wires. The semantics for the two binary control signals are fairly obvious. If just the **open** line is active, the fingers open. Similarly, if just the **close** line is activated, the hand grasps. However, let us specify that, for the special case in which both **open** and **close** are stimulated, the hand should stay in the same place.

open	close	finger action	"velocity"
0	0	don't care	undefined
0	1	squeeze together	+1
1	0	draw apart	-1
1	1	stay put	0

To see how this achieves the desired effect, suppose that the fingers are closing but the force is too high. Because the gripper is squeezing, initially **open** is false and **close** is true. However, when the CRUSH module kicks in it forces **open** to become true as well. According to the semantics defined above, this combination instructs the fingers to stay exactly where they are. Thus, the force regulation portion of the control works as desired. Now suppose that some other module decides to open the fingers by setting **open** true and **close** false. Since CRUSH only affects the **open** line and has already set it to the state desired by the other behavior, the new command essentially passes through unaltered. Thus CRUSH relinquishes control when the hand is commanded to move in the direction which naturally reduces the pressure. Assigning numbers to the velocity commands as shown above, we can see that partial suppression is actually a weak form of subtraction in this case.

3.4 Controlling the arm locally

Let us turn our attention to the behaviors controlling the arm. Figure 3-8 shows the complete set of 14 behaviors. Like the hand controller, this collection is broken into a number of levels of competence. Thanks to the hand controller, the robot can already grab things which happen to be between its fingers. We now extend the range of the robot so that it can grasp things that are further from the hand. However, instead of installing additional, more sophisticated grasping procedures, we build on previous levels. By reducing the problem to a situation the robot already knows how to handle, the pre-existing routines can take over and finish the job. For instance, the most primitive level of the arm controller, the **Local** level, tells the robot how to react to an object that is in close proximity to its hand. The robot lifts its hand until the top of the can is found, and then extends the hand slightly straight forward. This set of actions brings the can between the two fingers at which point the **Grip** level exactly knows what to do. The **Local** does not need to duplicate this skill or invent different ways of grasping things. Thus, although the various levels do not depend on each other's structure, they do depend on each other's functionality.

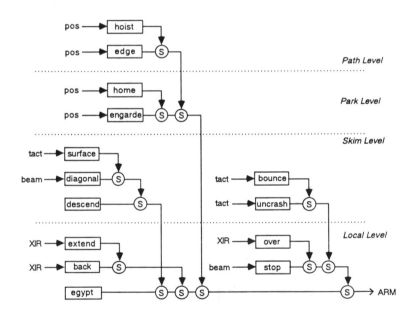

Figure 3-8. The control system for the arm.

The acquisition strategy embodied by the **Local** level is actually implemented by a collection of 3 separate behaviors. Each module recognizes a particular arm-object configuration based on readings obtained from the crossed infrared proximity sensors. For each of these configurations we specify a direction to move the hand. When active, the associated module continuously generates this command. For instance, the OVER module helps protect the hand by responding to things which fit its ungraspable object "model". Specifically, when both of the front IRs simultaneously detect something, OVER raises the hand to clear the hurdle. Normally, the EGYPT modules freezes the arm in whatever awkward angular configuration it happens to be in at the time. The new command produced overrides this zero velocity default. Notice that OVER *directly* drives the arm; it docs not just record a fact in some database for future reference. While other researchers have experimented with hand mounted optical sensors (e.g. [Balek and Kelley 85]), they have typically been used for object recognition rather than as a navigation tool.

While the OVER behavior safeguards the hand, it does not actively promote the collection of cans. For this, we make the robot's overall "plan" more sophisticated by adding the EXTEND behavior. If the IR sensors have seen something in the last several seconds, EXTEND drives the arm straight forward in an attempt to locate the object again. Normally this action would be a bad idea since the hand would simply plow obstacles over. Thus, as shown in the subsumption diagram (figure 3-8), to assure the hand's safety we give the OVER module a higher priority. However, since OVER and EXTEND have identical triggering conditions, we also allow EXTEND to *remain* active for a short while after its triggering stimulus has vanished (by using a monostable). EXTEND implicitly incorporates an estimate of the robot's dynamics so that this time constant moves the hand a distance equivalent to the farthest sensor range. The relative priorities of EXTEND and OVER then generate the basic snatching movement necessary for acquiring cans. OVER first raises the hand until the IR signal disappears, then EXTEND gains control and drives the hand forward to bring the sensed object between the robot's fingers.

This motion sequence works fine if the soda can is centered with respect to the hand. However, if the can is displaced to one side OVER will not recognize it as an obstacle to avoid ramming into. Therefore, continuing our enumeration of the possible "states" of the robot's hand, we add a new behavior, BACK, specifically for the situation in which there is a laterally offset obstacle. When only one of the two IR sensors is active, this module briefly drives the hand backward and down. Notice that in the subsumption diagram (figure 3-8) the BACK module is inferior to the EXTEND module. Since there is no case in which the sensor patterns for the two modules could both be satisfied at the same time, one would imagine that the ordering between them did not matter. However, recall that EXTEND tries to pursue recently seen objects even if they are no longer visible. Thus, if BACK were given priority over EXTEND the hand might reverse itself just as it was on the brink of grasping the soda can. This is especially likely since, as targets move in and out of sensor range, the two IRs seldom switch at the same instant.

We have now wired-in the basic acquisition procedure but the robot's performance needs to be tuned up in certain cases. In a subsumption-like architecture this is particularly easy to do -- we just add another module for the troublesome situation. Here the problem is that EXTEND often interferes with the grasp reflex. Although the fingers can move at 3" per second, they are usually up against their outer stops and must first overcome some residual servo error before moving. If the hand keeps travelling after the object is between the fingers, it may actually go beyond the object before the fingers have closed sufficiently. Therefore, we add the STOP behavior to freeze the hand for a while if the robot detects any activity in the finger beam sensor. If the module sees either a rising or falling (for the deposit phase) edge it locks the arm in the current configuration for several seconds. Using the edge rather than the signal itself keeps the hand from being frozen in place forever after something passes between the fingers.

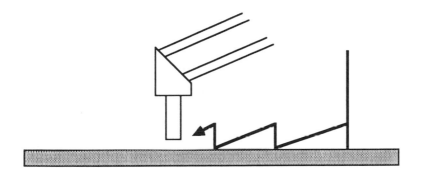

Figure 3-9. The **Skim** level contains a number of behaviors which together cause the hand to hop along the surface of tables.

The robot now has the capability of snatching nearby objects with its hand. The next level of competence, **Skim**, extends this capability by allowing the robot to acquire objects which are not initially perceived by the crossed IR. This new group of 5 behaviors basically contains an operational representation of the fact that objects often rest on surfaces. The control system's strategy is to first drive the arm straight down until it contacts some surface. The hand then recoils straight upward until it is a short distance above the surface. At this point, it extends forward while descending slowly in an attempt to make contact with the surface again. The overall effect is that the hand hops along the top of the table as shown in figure 3-9. Thus, the separate behaviors cooperate to find and explore a surface, hopefully bringing a can within range of the local IR sensors.

Again, we do not try to build a tactile model of the robot's surroundings and thcn plan a path through it. Rather, the overall pattern of actions is determined on the fly by the specific configuration of the environment.

The collective behavior of this level bears a passing similarity to the generalized damper control scheme [Lozano-Pérez, Mason, and Taylor 84; Erdmann 85]. In these systems an arm is given some velocity vector to follow. If the arm contacts a surface and is unable to move directly along the commanded direction, the velocity error is transformed into a force normal to the surface. The manipulator is still free to move tangentially along the surface provided the original vector was outside the surface's friction cone (pressing too directly into a surface causes the hand to stick). It is instructive to examine the use of this scheme when coupled with a fixed initial velocity vector. In particular, it has been shown that there are whole regions of space, called the task's "pre-image", from which the operation can be successfully completed with no additional high-level control. The manipulator travels through free space then slides along various surfaces to reach its goal. For our approach the whole area above and behind the soda can can be considered the task's pre-image. Although our system is not as elegant as the generalized damper control law, it can be made to work with the limited manipulator dexterity and sensory capabilities available to us.

The most basic module in our extended control system is DESCEND which simply drives the hand straight down in order to find a surface to explore. However, if the robot is already grasping an object this action must be modified somewhat. This is the function of the DIAGONAL module which overrides the more general purpose DESCEND module. If the robot pushes a can straight into a surface, it has to exert a considerable amount of force before the wrist switch will trigger and cause the can to be released. When this happens, the built-up strain on the arm is suddenly released and the robot "spikes" the can into the table top, like a football player who has just make a touchdown. When the hand is grasping a can, much less force is necessary if the wrist can instead be induced to roll slightly. Thus, DIAGONAL gives the hand a slight bit of forward velocity in addition to the primary downward component. The actual direction of travel is matched to the robot's rectangular workspace such that the hand starts in the inner, top corner and ends at the outer, bottom corner. In this way the robot still seeks a supporting surface but is better able to deposit the can without pounding it into the table.

Once some obstruction has been encountered, another behavior, BOUNCE, comes into play. This behavior prevents the robot from jamming its fingers through table tops by forcing the hand to recoil if either of the tip switches has been activated. As with other modules, we allow BOUNCE to remain active for a short while after its tactile triggering stimulus has disappeared. The length of this continuation, coupled with the dynamics of the arm servos, determines how far the hand will actually rise. We use this timed approach because the arm's position sensors are not accurate enough to let us build a closed loop system for the small distances we wish to travel (about 1/2"). With respect to the task, however, the precise height is of little concern. Whether the hand rises 0.2 inch or 0.7, it still has sufficient clearance to move laterally. Furthermore, since table tops usually occupy a particular region in our robot's workspace, the arm's speed only needs to be constant locally.

After the initial bounce, the SURFACE module prompts the robot to search along the top of the table. When any of the tactile sensors have been activated, this module directs the hand to go forward and slightly down for a while. Eventually, if the surface is more or less flat, the hand will descend far enough to activate the tip switches again and in turn retrigger BOUNCE. The length of each hop, therefore, is determined by height of the bounce segment combined with the commanded angle of descent. Actually, the downward component of the motion is not strictly necessary since DESCEND will take over once SURFACE times out. However, including this component causes the SURFACE module to continually verify its assumption that there is a surface close by.

Finally we have the UNCRASH module which performs a function similar to BOUNCE, but for vertically oriented surfaces. It detects when the hand has run forward into something by looking for situations in which the wrist senses a force but the fingers do not. In response UNCRASH causes the hand to simultaneously back away from the obstacle and to rise relative to it. Notice that if the robot continues to advance after UNCRASH terminates, the hand will prod the obstruction once again, but at a higher level. Thus, the hand not only hops across surfaces, it hops *up* them as well. This is similar to a spinal stepping reflex observed in cats. If the front of the foot contacts a rock in the cat's path, the creature will twitch its foreleg back and try placing it at a higher level. An identical strategy has been successfully used in a stair-climbing mobile robot [Hirose et al. 85] and as well as in one of Brooks's more recent robots [Brooks 89].

The sensory pattern which triggers UNCRASH can also occur when the robot has just placed a can on a surface. In such situation the wrist will actively sense a force, but the fingertips will remain inactive because they are not in contact with the table. It turns out that the same action is appropriate in this case as well. Even after DEPOSIT releases the can, there will be some accumulated servo error in the arm causing the hand to press against the top of the can. The upward component of UNCRASH's command helps alleviate this residual force. Meanwhile, the backward component causes the finger beam sensor to clear the back edge of the can. This is an important side effect because it prevents the robot from immediately regrasping the object it has just put down. The fact that UNCRASH works unaltered in this situation is not as coincidental as it might seem. Because the wrist cannot discriminate between upward and backward forces, the system moves to relieve both potential stresses when this sensor is activated. Furthermore, because the wrist was intended to be used during the exploratory phase, it was made sensitive to backward pressures. Its associated movement, therefore, locally reverses the normal grasping sequence and so naturally deactivates the finger beam sensor.

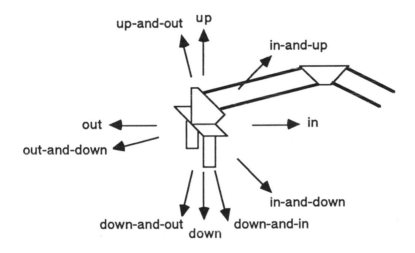

Figure 3-10. The hand is commanded to move in one of 10 relative directions.

Now that we have described the entire local grasping system, it is interesting to note that there are only a few characteristic movements executed by the system. We never need to use the full generality of the servos to drive the arm in some arbitrary direction. Just because the hand *can* be moved in any direction, we do not necessarily *have* to use all these directions. Restricting the arm to a small set of stereotyped motions makes calibration or learning of the manipulator's inverse kinematics easier since we only have to solve for these special cases. Figure 3-10 shows the 10 local directions of motion used for both local and global control of the arm. These motions are all in the vertical plane which is the arm's workspace and are the only directions required for proper operation of the arm.

3.5 Controlling the arm globally

So far we have discussed reactions which occur relative to the hand itself. The robot also has several behaviors which coordinate movement relative to various areas within its workspace. As figure 3-11 shows, there are a small number of special areas the robot knows about. These 8 separate bits of information are all the robot needs for normal operation; we do not really care about the hand's absolute x-y coordinates. As with the canonical directions of travel for the arm, limiting the robot to a few key items aids in the calibration, or possible learning, of the requisite spatial knowledge. A similar minimalist qualitative spatial representation has been proposed for a computer controlled prosthetic leg [Bekey and Tomovic 86].

The locations shown are typically used to signal transitions from one type of behavior to another. For instance, when travelling the robot folds the four links of the arm against each other causing its hand to be positioned directly in front of the main body. This special tucked in location is marked by the roughly 4" square zone marked AT-HOME. The corresponding one bit predicate serves as a termination condition for behaviors which retract the arm. A looser area, denoted IN-YARD, is useful for behaviors which want to know whether the arm has been safely parked yet. This is an example of behaviors communicating through the world. Finally, there are two extended boundary predicates, ABOVE-HOME and BEYOND-HOME, which simply help the robot determine the qualitative direction toward its parking location. Four more binary predicates delimit the borders of the rectangular logical workspace. The far edge is marked by BEYOND-SPACE and the top by ABOVE-SPACE while the inner edge has two boundaries associated with it. At the beginning of a grasp cycle the hand travels all the way to IN-SPACE to ensure that it will not miss the table top when descending. On the way back to home, however, the robot takes special care not to hit the edge of the table. Yet once the hand gets to the BEHIND-SPACE line, any motion toward home is safe and the more cautious approach is no longer needed.

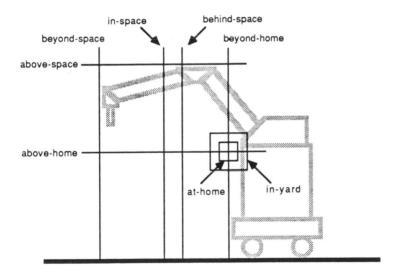

Figure 3-11. These 8 position predicates are the only spatial knowledge the system uses. AT-HOME and IN-YARD denote roughly rectangular regions whereas the other 6 predicates mark useful half-planes. For instance, BEHIND-SPACE is true if the hand is anywhere to the right of the vertical boundary line shown in the diagram.

The first level of the global control system, **Park,** is concerned primarily with the home location. The ENGARDE module in this level controls the deployment of the arm after it has been parked awhile. Like a swordsman, this module starts by moving the hand from the tucked travelling position up to near the top of its workspace and then out towards the inner edge. This in turn helps the DESCEND behavior find a suitable surface. If the hand started directly from its parking spot, the robot would fail to explore the top half of its workspace. Furthermore, even if the table is lower than the parking location, the hand will miss the surface altogether unless the robot is positioned exactly against the table. ENGARDE's transfer function is relatively transparent: it tells the hand to go diagonally upward until it achieves sufficient height and then to proceed straight outward. The initial direction of travel is closely matched to the direction between home and the top innermost edge of the workspace. However, to ensure that the hand is placed over the table, ENGARDE only stops when it reaches the further of the two inner workspace boundaries. At this point ENGARDE shuts down and lets other behaviors take control. In addition, the module has a short refractory period (implemented by a monostable). This prevents ENGARDE from being immediately reactivated if, say, DESCEND accidentally crosses back over the workspace boundary.

Besides just finding and grabbing cans, the robot must also bring them back closer to its body in order to transport them safely to another location. The **Park** level contains another behavior, HOME, which takes care of this. HOME instructs the hand to move (using a relatively complicated transfer function) on one of the four main diagonals to reduce its relative offset from home. The decision to retract is based on the special motion predicate which is updated every 5 seconds or so. If the hand has failed to move at least 1/2" in this interval, the variable is set true. After a valid stop is detected, we allow the arm about a minute to reach home. If it has not succeeded in this interval, HOME automatically gives up. Usually, however, the module successfully parks the arm and then holds it at home for the remainder of the allotted time. Unfortunately, satisfaction clauses were an after-thought and were not included in HOME's applicability predicate. Thus, its internal monostable does not get deliberately reset when the hand reaches the parking position. If the hand happens to be very close to start with, it may end up locked at the home position for a long (but finite) amount of time.

The HOME module was originally installed simply as a protective measure to cope with the cases in which the robot can no longer make meaningful progress. For instance, the arm might reach the edge of its working envelope, or become stuck on some obstruction. In these circumstances, a reasonable thing to do is to retract the arm. Observing that in both cases the arm ceases to move, we designed the HOME behavior to trigger whenever the hand stops at a non-home location. When this happens, HOME attempts to bring the hand back to the special parking location. In a sense, motion of the hand is used as a progress indicator. If the hand has stopped, the current set of behaviors is no longer operating effectively and should be replaced by a different strategy. Thus, although the entire groping procedure relies only on local sensory information, we can use a single piece of global information to give it an overall perspective.

This motion predicate can also be used as signal between behaviors. Although we do not allow direct connections between modules, one module often alters the relation between the robot and its environment to cause another module to become more effective. For instance, skimming along the table hopefully brings a can within range of the IR sensors thus allowing the local grasping routine to run. Here, we use this technique more directly. Recall that we *deliberately* stop the hand when the finger beam is activated in order to allow the fingers time to open or close. This in turn naturally triggers the HOME behavior and thereby brings the the arm back after the robot has grasped or ungrasped an object. Thus we do not need a new, separate behavior to haul cans in. We can instead simply tap into an existing reflex by setting up its local environmental activation pattern.

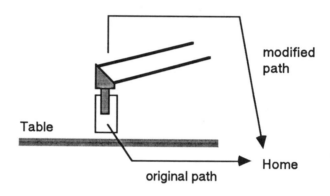
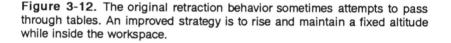

Figure 3-12. The original retraction behavior sometimes attempts to pass through tables. An improved strategy is to rise and maintain a fixed altitude while inside the workspace.

The next level, **Path**, refines the basic retraction strategy embodied in HOME. This new level encodes the fact that cans are often found amidst clutter. The primary module, HOIST, is activated by essentially the same environmental conditions as HOME, but it does not head directly toward the parking location. Rather, it first lifts the can almost to the top of the workspace to clear any other objects on the table, then brings it straight back to the innermost edge. As shown in figure 3-12, going directly home is often a poor choice of trajectories. This is especially true when the robot has been exploring a surface at a height above that of the parking location. If the hand is empty, it will bounce backward along the table until it reaches the edge and then proceed straight home. If the robot is grasping something, however, the hand will jam it against the table top causing the DEPOSIT and UNCRASH behaviors to release the object. The "up then back" strategy used by HOIST eliminates this difficulty. Yet HOIST does not completely subsume HOME; once the hand leaves the workspace HOIST no longer generates any outputs and allows HOME to finish retracting the arm. This is another case where a more specific behavior (HOIST) suppresses a general purpose behavior (HOME).

The other behavior in this level, EDGE, stops the hand (and hence causes retraction) whenever the hand goes beyond the far edge of the inscribed rectangular workspace. While the hand can often reach beyond these limits, it can not always rise straight up all the way to the top of the workspace. Thus, EDGE makes sure the hand does not go beyond the point where the HOIST strategy will operate properly. The limit stop imposed by EDGE has the added advantage that it also keeps the arm in the "elbow up" configuration. Just beyond the lower outside corner of the workspace the arm is at its fullest extension so there is only one solution for the joint angles. Unfortunately when leaving this point the controller sometimes inverts the elbow. Because the parking location can only be reached with the elbow up, the robot may not be able to return home without passing through another singularity. The EDGE module therefore also serves to prevent this dilemma by never letting the arm reach too far. This is almost the converse of the relations between modules that have been previously discussed. Instead of actively reorganizing the world so that a particular behavior will start to work, EDGE ensures that the world does not change so much that the retraction behaviors will cease to work.

3.6 Controlling the base

The system described so far can retrieve cans from a wide range of heights and at a variety of distances from the robot's body. However, the arm operates only in a single vertical plane. For successful operation the can must be aligned with this plane to within the tolerance of the gripper (about 0.75" either way). The final level of the arm controller, **Swivel**, loosens this restriction by using information from the hand IRs to turn the entire robot and help center the can. As shown back in figure 3-5, the hand's crossed infrared sensors respond over a region nearly three times the width of the "grab zone". We can use, for instance, the activation of just the left sensor to tell us that the can is offset to the right. However, the response regions for the two sensors cross over the centerline when the hand is about 1" off the surface of the table (compare middle and bottom panels). Thus, we need to keep the hand reasonably high to avoid control inversion which would send the hand in exactly the wrong direction. Normally the hopping motion produced by the BOUNCE module is sufficient to guarantee this.

Due to a mechanical linkage, the arm always faces in the same direction as the robot's wheels. To perform the necessary sideways shift we rotate the base in place. Most of the time, however, we do not want the base to move at all. Thus, we have created two default behaviors, FREEZE and LIMP, which prohibit translation and rotation respectively. We use two modules because the base treats orientation and velocity as independent resources. The control signals to the base can take on one of three discrete values: increase, decrease, or don't change. The modules shown here continuous generate the appropriate "don't change" messages for their degree of freedom.

The TWIST module controls the actual reorientation of the base. As shown in figure 3-13, to do this it needs only to affect the base's rotary degree of freedom. TWIST's applicability predicate waits until it is sure that it has seen an object on one side only. This long delay prevents the module from making the costly mistake of turning. It also has the fortuitous effect of giving the BACK module time to stop the forward advance of the hand. Once TWIST has been triggered, we gate the module's transfer function to the output for only a fixed amount of time. When an appropriate sensor pattern is detected, we use the rising edge of the recognition predicate to set the module's mode memory monostable. TWIST's transfer function then attempts to drive the base of the robot so that both sensors see the can. If, in the course of this movement, the can is lost (seen by neither sensor) the robot stops. However, the maximum turn the module can make is governed by the time constant of its internal monostable.

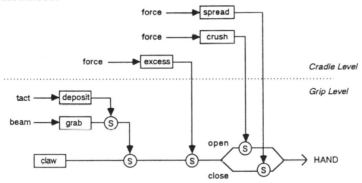

Figure 3-13. The arm is reoriented by the TWIST and QUENCH modules which override the default behavior for base's rotational degree of freedom.

In general, control of the base is very difficult for a variety of mechanical reasons. For instance, TWIST does not use position control for rotation because the necessary angular resolution is not available. Although the position encoders on the base can sense very small rotations, the arm is not rigidly attached to the base. The whole torso of the robot can twist a bit and the arm itself has a tendency to wobble when fully extended. Thus the base's sensors do not accurately reflect the position of the hand. Even with timed commands, the rotations required are so small that if the base is active for long it will overshoot the centerline. Typically, we want the hand to move on the order of one inch at a radius of 20 inches (i.e. roughly a 3 degree rotation). Using the fastest base command update rate and turning at the slowest reliable speed we compute that the minimum achievable turn is about 5 degrees. Fortunately this is mitigated by the fact that the base has some initial backlash to overcome and it sticks a little when it starts to turn. In addition, the upper body has a large rotational inertia which limits the robot's acceleration and deceleration. However, the performance of this system could be improved significantly if the base incorporated a real closed-loop velocity servo and if its command language overhead was reduced.

Although the moves made by TWIST are small, sometimes the robot still oscillates around the centerline. When the left IR sees something, TWIST tells the base to turn to the right which eventually causes the right IR to come on. Normally, this signal would cause the robot to start turning back to the left. To prevent this motion we add another module, QUENCH, which provides some hysteresis for the system. QUENCH watches for the direction of asymmetry to change and, when this happens, locks the base in place for a while. The change in direction is detected by using a pair of monostables to remember whether either IR has been on recently. If both IRs have been on recently (not necessarily at the same time) QUENCH steps in. This essentially allows the robot to turn in only one direction and thus prevents any backward rotation. Notice that QUENCH can only be deactivated by a long interval in which *neither* IR is active. This serendipitous feature is useful because it keeps the robot form jerking sideways due to asynchronous extinction of the IR signals when the hand rises to the top of the can.

3.7 Experiments

The system described above has been implemented on our multiprocessor and works well in practice. The first thing we investigated was the true efficacy of the **Swivel** level. In a series of 40 consecutive trials the robot achieved a 90% overall success rate. A trial was considered successful if the robot eventually achieved a reliable grasp on the can. In these experiments the can was placed in the center of the arm's workspace and displaced slightly from the hand's normal line of travel. Specifically, we tested left and right offsets of 0.5" and 1.5". The robot's actions were similar for both values, even though the smaller offset was within the robot's natural grab zone. On 25% of the successful trials, the robot overcompensated on the first turn and had to perform a second turn to properly orient the hand. The 4 failures in our tests were caused when the hand did not rise high enough to pass over the can. Although the can was knocked over, in most cases the hand was properly centered. Thus, we can count on the hand to grab any can whose center lies within 1.5" of its centerline.

Next we investigated the reliability of the local and global control behaviors used by the arm. Figure 3-14 shows a composite of the trajectories produced by the system on 10 consecutives runs. The data was obtained by video-taping the arm in action and then manually plotting the position of the fingertips during slow-speed playback. A complete run takes approximately 2 minutes. On 30 consecutive trials, evenly distributed between the three environments shown in this section, the hand achieved a 90% overall success rate. Notice that this was accomplished despite rather poor manipulator control (none of the path segments are straight) and with no prior model of the environment.

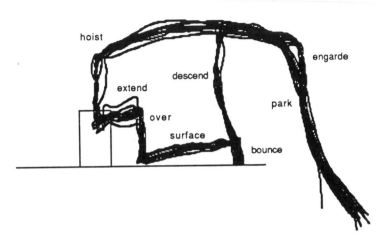

Figure 3-14. The result of 10 consecutive runs on the nominal environment. The lines show the path followed by the tips of the fingers.

The behaviors responsible for various phases of the trajectory are noted on the plot in figure 3-14. In the beginning ENGARDE deploys the arm to the top edge of the workspace. Then DESCEND takes over and drives the hand down until the fingers contact the table top. At this point BOUNCE is activated to lift the hand slightly, and SURFACE kicks in to urge the hand along the table. When the hand gets close to the can, OVER lifts it to a suitable grasping height while EXTEND continues to propel it forward. Eventually the finger beam is activated which stops the hand and causes GRAB to close the fingers. Seeing that the hand is stopped, both HOME and HOIST become active. Initially HOIST wins the arbitration, lifts the can up to the top of the workspace, and brings it in toward the robot's body. After clearing the work area HOIST relinquishes control and HOME takes over to bring the hand the rest of the way to its parking position.

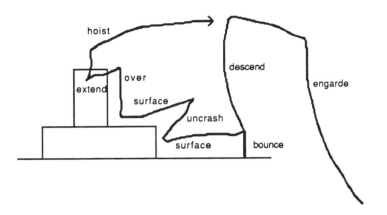

Figure 3-15. The same control system can retrieve a can from a pedestal. Although the IR sensors did not notice the rise, the ensuing collision was detected by the wrist switch.

Because it has very few expectations about the world, the same control system can be used in many different environments. Figure 3-15 shows the trace of an actual run in which the can was not directly on the surface of the table. The arm starts out in the same way as before, finds the table, and starts groping along the surface. At a certain point, however, the crossed IRs detect the book which is supporting the can, and cause the hand to rise to a suitable height to grip the *book*. EXTEND then drives the hand forward trying to place the book between the robot's fingers. This attempt is doomed to failure and soon causes the wrist switch to be activated as the hand rams up against the side of the book. When this happens UNCRASH takes over and raises the hand still further. This time the hand successfully clears the book and goes on to grab the can in the standard fashion. A similar pattern of behavior has been observed when the hand just misses the edge of the table during the initial descent phase. The crossed IRs notice the table and then, the the aid of the wrist switch, the hand navigates back onto the top surface.

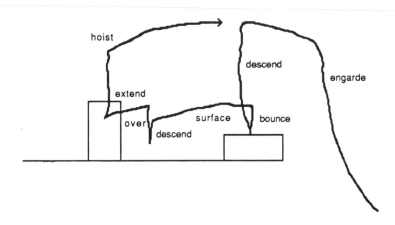

Figure 3-16. The arm can also collect cans from behind a barrier. After exploring unsuccessfully for a short while, SURFACE drops out and lets DESCEND take over.

Another example of the controller's flexibility is exhibited in figure 3-16. Once again this a recording of the actual operation of the arm, but this time the can has been hidden behind a barrier. The robot starts by locating the top of surface of the book in the normal manner. However, as it is bouncing across the surface the robot notices that the tip switches have been inactive for an abnormally long time. The monostable inside the SURFACE module thus times-out and control reverts to the DESCEND behavior. As it happens, while the hand is adjusting itself to the new terrain height, the crossed IRs pick up the can and the normal grasping cycle is initiated. If the can had been further behind the edge of the book, the hand would have instead descended all the way to the surface of the table and started skimming across it in search of a can.

This example points out the danger of incorporating excessive state into the system. When the fingers initially made contact with the book, the robot assumed it had found the correct height at which to search. It implicitly models the environment as a single flat surface with only this one parameter. Fortunately it stored the height information by activating SURFACE's monostable, a piece of *self-decaying* state. Like a watchdog timer, the period of the monostable was set so that, for a truly flat surface, it would normally be retriggered before it timed out. When this does not occur (no contact is made) the robot automatically "forgets" about the original event. Thus, the robot never acts for long on outdated information and is free to adapt its behavior to whatever circumstances are present at the moment.

While our system does not *construct* representations or *build* plans, it does use them in the course of activity. As demonstrated by the experiments, there are obviously many partial representations and sketchy plans embedded in the control system. Yet these are present only at "compile time". That is, they are part of the creature's construction rather than something subject to change over time. For instance, when the hand gropes around for a can it is implicitly using a representation of a table. Part of this is representation is contained in the way the fingertip sensors were installed, part is in the standard directions of movement of the arm, and part in three independent control agents. As we have seen, these agents also contain a rudimentary plan about how to find and search across a table-like surface. One agent "knows" that tables are contacted by going down. Another agent "knows" that the hand must not contact the table if it wants to move. The third agent "knows" that the table can be followed by extending the arm outward. However, this plan was built into the creature as a collection of instinctive behaviors, rather than being deduced on the fly by some conscious, introspective reasoning facility. The robot's apparent ability to dynamically plan trajectories is actually derived from the sequence in which environmental conditions allow these behaviors to be triggered.

Chapter 4

Vision

With the development of the arm control system in the last chapter, our robot is now capable of finding and grasping cans that lie anywhere within the arm's workspace and that are roughly aligned with the plane of the arm. But what about cans that remain beyond the range of the local sensors? This is the purpose of the vision system: to locate suitable objects from a distance. With this new capability, the robot's collection zone can be expanded to include objects that are within 30 degrees of the robot's direction of travel and up to 6 feet away.

In our system we do not adhere to the paradigm of model-based matching followed by a transformation from image coordinates into joint angles. Rather, with our approach it is only necessary to map the edges of the arm's effective workspace into visual coordinates. To grab a can, the robot simply moves itself so that the image of the object lies in the image of the workspace, then releases the arm. The rest happens automatically thanks to the collective competence of the arm behaviors. In this way, the process of localizing the can is shared between the vision system and the arm controller.

This is the key idea of this chapter: that the complexity of hand-eye coordination problem can be greatly reduced by relying on the behavior of the arm controller. While many other researchers have found ways to learn the full mapping from image coordinates to joint angles (e.g [Kuperstein 87]), none of these systems is as simple as ours. The advantage of having a complete mapping is that it would allow us to move the hand more directly toward the can. Yet, for our task all that matters is whether the robot eventually grasps the can, not how efficiently it does this. Furthermore, even if we could approach the can directly, we would still require many of the local obstacle avoidance and fine positioning strategies of the arm controller. Since these behaviors were already present, we were able to take advantage of them to simplify the task of adding vision to our robot.

Many sorts of coordinate transformations can be simplified in a similar way. Suppose we had an additional sensor that could discriminate red objects from green objects. To grab red cans, the robot would first center a can shape and then see if a red object appeared at the corresponding canonical location in the color image. It is not necessary to derive a dense correspondence mapping from one set of image coordinates to another. We can instead map just a small region of correspondence and then arrange to bring potentially interesting objects into this focus of attention.

The rest of this chapter describes the actual mechanisms used for finding cans and positioning the robot. We first describe the nature of our range sensor and the hardware used for processing the images. Next, we discuss the properties of real range images and show how these are incorporated in our hard-wired can recognition algorithm. We then describe how the recognition information is used to drive the two degrees of freedom of our robot. Finally, we report on the empirical performance of the recognition algorithm and the proficiency of the robot body alignment system.

4.1 Hardware

Before discussing how the robot locates cans it is first necessary to understand how the robot perceives the world. To achieve good discrimination and a wide field of view we need a sensor with high resolution. For this reason we decided to use an optical ranging system. While there are several systems commercially available, we found that none were suitable for our application. They were either too large, required too much power, or were too expensive. Thus, we chose to construct our own laser ranging system. The system is small enough to fit onboard the robot and, since it only consumes 30 watts of power, can be run off batteries. The actual sensor is shown in Figure 4-1.

Our range finder is based on a triangulation scheme (cf. [Essenmacher et al. 88]) rather than time-of-flight [Lewis and Johnston 77] or phase detection [Miller and Wagner 87] as some other systems. The large vertical column is a 7mw Melles-Griot helium-neon laser. At the top of this we have mounted a cylindrical lens to spread the beam into a 90 degree wide sheet of light (see figure 4-2 left). The power density of the laser radiation is approximately equivalent to that emitted by three 100W incandescent bulbs and so is bright, but safe. This sheet is then reflected off a movable mirror mounted directly above the laser. A low-inertia Densei stepper motor is used to ratchet the horizontally oriented plane of laser light up and down in front of the robot. A complete sweep takes 1.1 seconds and covers almost 60 degrees.

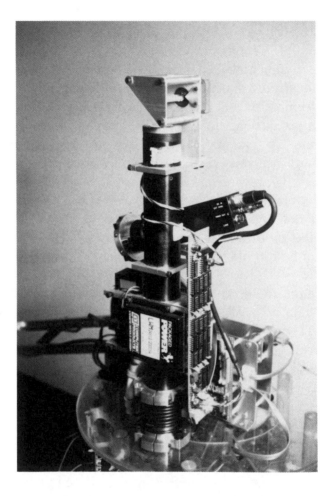

Figure 4-1. The laser light striper and its camera are mounted on the top of the robot. This sensor helps find cans in an area 60 degrees wide by 6 feet deep.

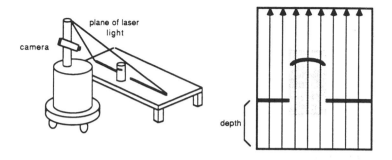

Figure 4-2. The robot locates objects by sweeping a plane of laser light up and down (left). This is observed by an offset camera whose scan lines are oriented vertically. We compute depth by measuring the distance between the laser line and the edge of the picture (right).

Lower down on the laser tube is a Pulnix TM-540 CCD camera (AGC disabled) which is fitted with a wide angle lens to give it a roughly 60 by 75 degree field of view. The camera is tilted downwards 20 degrees and rolled on its side so that its normally "horizontal" scan lines run from bottom to top. This re-orientation of the camera is the key to the operation of the system. At the beginning of each camera scan line we start an 8 bit timer which then counts the pixels along the scan line until it detects the laser beam (a similar idea was used in [Popplestone et al. 75]). The further from the beginning of the scan line the laser stripe is detected, the closer the object is to the camera (see figure 4-2 right). We perform this measurement for each line of the CCD image to get 256 disparity readings per picture. All this happens during the odd field of the NTSC signal. During the even field we increment the position of the deflection mirror and let it settle. A full range image requires stepping the sheet of laser light through 32 such discrete deflections.

Since we use an intensity threshold to detect the laser stripe in an image, we employ several methods to improve the rejection of ambient illumination. First, we noticed that bright objects in the background, such as windows and pieces of paper, often have a large spatial extent. Thus, we use special digital circuitry to look just for skinny lines in the image. In addition, to emphasize the red light from the laser, the camera is equipped with a broadband interference filter from Oriel. This device is centered at 650nm and has a 70nm bandwidth. While narrower filters were available for the particular wavelength of interest (632.8nm), the center frequency of all interference filters shifts as the angle of the incoming radiation increases (see figure 4-3 left). By using a wide filter we achieve a relatively flat transfer function over the region of interest (figure 4-3 right) while still boosting the signal to noise ratio. Other researchers (e.g. [Pipitone and Marshall 83]) have avoided the incident angle problem by using a narrow filter and sweeping both the laser and the detector. This trades speed and simplicity for better sensitivity.

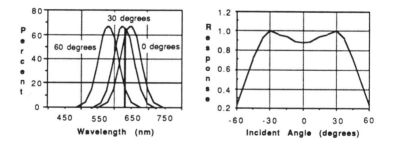

Figure 4-3. A filter is placed in front of the camera to aid in locating the laser stripe. The pass band shifts relative to the laser frequency (dark line on left) depending on the direction of the incoming light ray. This makes the center of the image dimmer than the edges (right).

We have also designed a special set of processors to interpret the data from the light striper in real-time. Each Line-Oriented Vision Processor (LOVP) board consists of an 8 bit micro-processor (a Hitachi 6301), 2K of RAM, a high-speed data transceiver, and a general-purpose parallel port. Because these boards are very small, 3 by 4 inches, and do not require much power, about 1W each, all the necessary processors can be mounted onboard. This is an important consideration for mobile robots because external cables invariably get tangled. Similarly, the telemetry necessary to process images offboard is generally infeasible due to the physically cluttered and electrically noisy indoor environment.

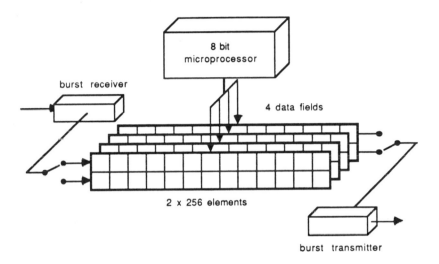

Figure 4-4. The Line-Oriented Vision Processors (LOVPs) are based on a standard 8 bit microprocessor. Each board only has access to the two most recent horizontal lines of the depth map. Special hardware automatically scrolls these lines into and out of memory.

The LOVPs communicate with each other using short bursts of image data. Each 1/30 of a second the light-striper generates a set of 256 disparity measurements corresponding to the wiggles and jumps observable in the laser beam. This set of numbers is then compressed into a 0.5ms burst and sent to the first LOVP. When a LOVP receives data, special hardware stores it directly into the onboard RAM chip. The RAM is divided into two conceptual chunks: old-space and new-space. In new-space we have the most recent line of data plus three 256 byte sections worth of "scratch-pad" used for storing temporary results. Similarly, in old-space we have the second most recent line and its associated scratch-pad sections. When a new burst of data is received, the current new-space is renamed old-space, and the incoming data is written to new-space. After the incoming burst finishes, the receiving LOVP generates its own burst and sends all of the current old-space data to the next processor(s) in line. This produces an auto-pipelined structure: if the processor does nothing to the data in RAM, the LOVP acts as a simple delay unit. Normally, however, the microprocessor actively manipulates the available data between bursts. In this 32ms interval approximately 40 instructions per pixel can be executed.

The LOVPs can be connected in a chain or can fan out into a tree. They cannot, however, recombine later on. That is, a LOVP can talk to any number of other LOVPs but can only listen to one. This is not a serious restriction because we can use the scratchpad sections to perform limited recombination of data. For instance, the first LOVP could put its result into a scratchpad section and then transmit the original raw data and the newly processed data to the following LOVP during the next burst. This LOVP could then perform its processing on the raw data and store the result in a different scratchpad section. Finally, the third processor in the chain could take the two scratchpad sections and combined them to obtain a single result.

The LOVPs are well matched to many mobile robot applications. First of all, they process the data as fast as our light-striper can generate it. Second, an indefinite number of LOVPs can be added to the system without slowing it down. Each extra processor does, however, add a 1/30 of a second latency to the overall system. The LOVPs are also simple and cheap. No large and power hungry items such as frame-grabbers, 32 bit floating point vector processors, or hardware convolvers are required. Furthermore, the basic scan line sampling scheme can be extended to process gray level images as well. As shown in figure 4-5, any front end which yields one data point per scan line can be used. Here we average the video signal over a small interval (white boxes) and then send it to an ordinary A-to-D converter. Examining a single horizontal slice across the image gives information which can be used for motion or stereo vision [Brooks, Flynn, and Marill 87]. A coarse two dimensional image can be used to determine the orientation of ceiling lights as a navigational reference, or to follow objects and paths [Horswill and Brooks 88].

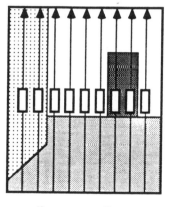
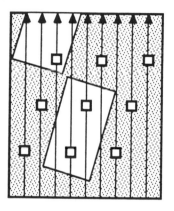

linear sampling   area sampling

Figure 4-5. The same line-oriented processing architecture can be used with gray-level images. The open rectangles indicate the sample points used in each image.

4.2 Image processing

The goal of our robot's vision system is to identify and report the location of cans within its field of view. The LOVP network sits processing depth images, one per second, and delivers to the subsumption architecture messages like: "I didn't see a soda can, I didn't see a soda can, I see a soda can at (110, 15, 60), I see a soda can at (115, 14, 62), I didn't see a soda can, ..." The numbers refer to the image coordinates of the can's bottom and a typical raw disparity value for the object. This is all the information currently reported; no confidence limits, alternative targets, size estimates, or shape descriptors are produced.

Yet, to process the real visual data adequately, we must first be familiar with the properties of the images actually produced by our light striper. Figure 4-6 shows one such image. We have stretched everything by a factor of 8 in the vertical dimension to restore the true aspect ratio. The gray levels represent the raw disparity measured; they have not been corrected for depth. That is, the computer has shaded each pixel based on the number returned by the front end of the light striper. This number corresponds to the measured distance from the laser line to the edge of the picture. Still, in general, the darker an object appears, the closer it is to the robot. This picture is the robot's impression of a soda can sitting on a bench top. There are various objects under the table, and a white wall behind it which is blocked at the bottom by other objects on the table. The lump in the lower right corner is not part of the exterior scene at all, but is the top of the robot's hand.

Notice that large areas of the image are pure white. These are places where the light striper received no returns. Since the position of the laser plane is detected by thresholding, if the stripe in the image is not sufficiently bright the robot will ignore it. This can happen if the object in the beam is too far away, has a low albedo (i.e. dark colored), is highly specular (i.e. shiny), or is oriented at a steep angle relative to the camera's line of sight. Note that some objects have narrow spikes on their surfaces. This occurs when the reflected beam is right at the detection threshold. Any slight variation of the reflecting surface causes pixels to erratically appear or drop out. The vertical streaks on the right side are caused by a different phenomenon. On each video line the light striper records the disparity of the first narrow bright spot it sees as it scans from the bottom to the top. Unfortunately, illumination highlights fit this description and yield the same disparity irregardless of the declination of the laser. Here the highlights are from the aluminum enclosure of the robot's own hand. We have since liberally applied electrical tape to the hand to help mask these bright spots.

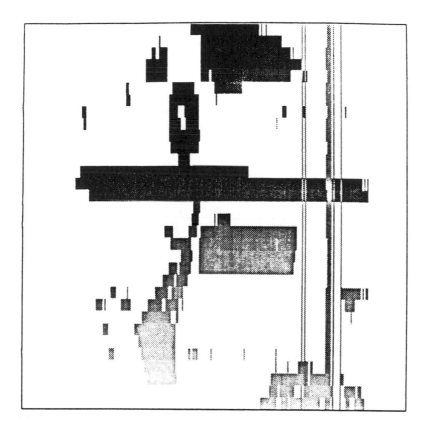

Figure 4-6. This is an actual range image produced by our laser striper. It shows a can on cluttered bench top. The streaks on the right are due to localized specular reflections.

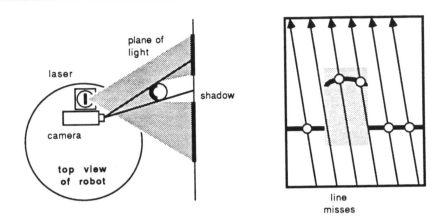

Figure 4-7. Vertical edges often have an adjacent no response region. Because the camera is offset from the laser there is a shadow on one side (left). If the scan lines are tipped slightly, a similar effect occurs on the other side (right).

Another interesting property of this image is that the soda can is relatively isolated. Except for the bottom, all of its boundaries are marked by a transition from true depth data to no depth data. In fact, in examining numerous images we have found that there are <u>very few depth discontinuities</u>. This is due to a number of factors. Flat horizontal surfaces typically do not show up because they are nearly parallel to the incoming beam. In addition, table tops are often have a glossy coating which does not produce much diffuse Lambertian scattering. Vertical edges are usually outlined by a no response region due to several geometric effects. Very seldom are objects with square edges aligned perfectly perpendicular to the camera. This means that there is at least one highly slanted surface (two for a can) which does not generate strong reflections. Also, as shown in the left half of figure 4-7, the sensor's camera is mounted slightly to the side of the laser. This leads to a shadowing effect on one edge of objects. A fortuitous misalignment of the camera can cause a similar narrow isolation band on the other side of objects as well. As shown on the right side of figure 4-7, if the scan lines are crooked they may pass between two sections of the laser stripe without registering anything.

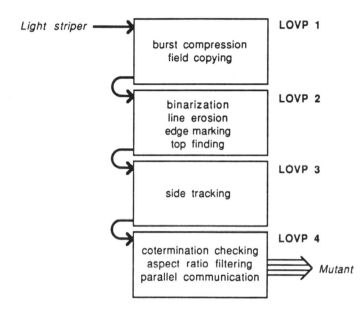

Figure 4-8. The robot uses a chain of 5 line-oriented vision processors to find cans. The first LOVP receives data directly from the light striper, while the last LOVP is interfaced to the subsumption architecture controller via a "mutant".

With these phenomena in mind, we designed a chain of 5 LOVP processors to help the robot find cans. This can detector is the only part of the system that would have to be changed if we wanted the robot to collect something else; the robot's orientation, manipulation, and navigation routines could all be re-used exactly as they are. Figure 4-8 shows a block diagram of the particular can-finder we use. The function of each LOVP is summarized and the interconnections between them are illustrated. Basically, the robot is looking for "staple" shaped boundaries (see the last panel of Figure 4-11). The robot considers an object to be a can if it has a flat top, and two fairly straight sides of equal length.

The first step of the actual visual processing is to convert the disparity map into a binary image. Any pixel for which a true depth reading obtained is marked as a one, all the others are zeroed. This simplification is possible because we do not use any depth information in the recognition phase of can finding. Next, to combat streaking, we eliminate any features narrower than 6 pixels. This should not remove any prospective cans because, even at a distance of six feet, they subtend at least 8 pixels. Next, as shown in figure 4-9, we look at the pixel above and to the left of the current pixel to determine whether an edge exists. We also classify any edges found as one of four types: left, right, top, or bottom. All this computation is done by the first LOVP in the chain.

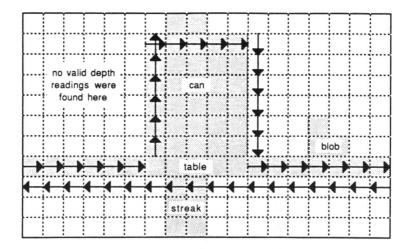

Figure 4-9. The first stage of processing yields a binary image in which all real depth readings have been replaced by ones. The system then eliminates any narrow noise features and marks the direction of the remaining edges.

Next, as shown in figure 4-10, we look for possible tops of cans. These consist of an upper left corner (a pixel marked as both a left and top edge), followed by a series of top edge fragments, and terminated by a right edge pixel. Since we only use horizontal and vertical edges, we consider objects to be distinct if they are not 4-connected. In addition, we require that the length of a proposed top be less than 40 pixels. This threshold is fixed and is not varied according to the computed depth of an object. It corresponds to the apparent size of a can placed 15 inches in front of the sensor. The top and edge maps are then passed to the second LOVP which looks for the sides of cans. These are chains of edge fragments which are more or less straight. In the actual code the maximum local dent allowed is 4 pixels, and the biggest bump is 4 pixels.

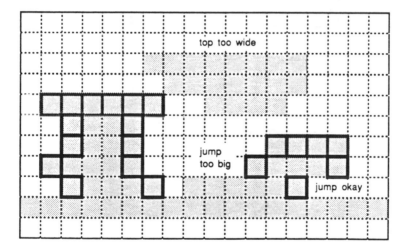

Figure 4-10. The system next looks for medium length horizontal boundaries (dark gray). It marks these as possible can tops and attempts to find approximately vertical adjoined sides (black). If the sides terminate at the same height, a can has been found.

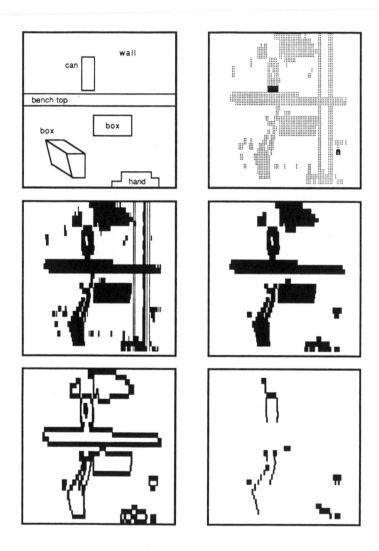

Figure 4-11. The upper left shows the composition of the scene. To the right, the depth map is binarized and narrow features are removed. At the bottom left we show the tops and sides found. Finally, the original image is marked with the best candidate (single underline).

Finally, the side map is passed to the last LOVP which finds the end of each chain. If two matching sides terminate at the same height in the image, the robot considers this convincing evidence for the existence of a can at that location. The LOVP then records the coordinates of the center of this can's bottom as well as its aspect ratio (length to width ratio). In addition, it averages the disparity values at the two side terminations to provide a representative depth measure for the can. As the image is scrolled through, the LOVP keeps track of the candidate whose true aspect ratio is closest to 2.7. The panel in the lower right of figure 4-11 shows the can selected by the processor (single dark bar) and one other possibility which was rejected (double bar). Once the complete image has been scanned in, the plane of laser light reverses its direction and starts sweeping upward. During this period the LOVP uses its parallel port to pass the parameters of the best can to an associated subsumption architecture processor which then makes the information available to the rest of the system.

4.3 Visual guidance

The front-end LOVPs deliver a map with the best can marked. This judgement, however, is based primarily on the shape of the object not its size. We did this to allow the LOVPs to work with a simple scale-invariant can description. Yet, the proper metric information could greatly reduce the system's mis-identification rate. Organizing the acquisition system as a collection of behaviors lets us add this in after the fact. In our application, we divide the robot alignment task into two separate operations. One behavior rotates the robot to bring the image of the can into the image of the workspace. Another behavior then uses the depth of the selected object to bring the robot within grasping distance. This approach behavior lets us dynamically weed out improperly sized objects. Recall that, to be perceived as the top of a can, a horizontal edge had to be less than a certain maximum width. Imagine that the robot erroneously latches on to some oversized object such as a printer. As the robot tries to approach to the correct grasping distance, the image of this object will grow in size. Long before the robot actually reaches the object, its visual width will exceed the specified limit and the robot will lose interest in it. So, not only can we use the behaviors of the arm to simplify the alignment process, we can also use the alignment behaviors themselves to simplify our model of cans.

Although not as important as our semi-procedural can model, the implementation of the alignment behaviors is also interesting. To convert the location of the can into a motion command for the robot we use something called a "space table". This scheme is adapted from the early work of Arbib [Arbib 81] and is a very flexible approach for dealing with images. After the LOVPs finish their work the robot has a relative "can-ness" map of the world in image coordinates. We view this entire transformed image as a big control table in which each pixel suggests some motion for the robot's body. The maximum of the relative "can-ness" map is then used to select a table entry. In our system only one pixel is ever marked, but one could imagine extending the approach to multiple targets with varying strengths. The particular control law associated with each component of the robot's visual control system is embodied in one of these "space tables".

For instance, the ALIGN module implements a feedback loop in image space. Its job is to determine whether the can is to the left or right of the workspace. It does this by assigning certain regions of the image to a left turn operator, and other regions to a right turn operator. This partitioning is based on the coordinates of the arm's workspace in the visual field. If the can appears to the left of this area, the robot should turn to the left and vice versa. If the selected object appears inside the workspace, no motion is necessary. On each scan of the light striper, ALIGN then activates the operator whose region includes the object selected by the LOVP image processors. This in turn emits a brief rotation pulse to bring the robot into line with the can. The length of each ballistic rotation is determined by the ALIGN module's monostable. Since these motions are small, the robot typically undershoots its objective and several cycles are required to properly orient the arm. Of course, if we had access to the base encoder values we could terminate the movement more intelligently. Still, adequate performance is obtained with just this simple scheme.

The other main control module is APPROACH. This compares the depth of the selected object against the near and far boundaries of the workspace. The values for the workspace were extracted from several images of a vertical post placed in the appropriate region. It uses the result of this comparison to determine whether to drive the robot forward or back. Like ALIGN, APPROACH uses the image coordinates of the most can-like object to select an entry of its control table. However, instead of a being a matrix of fixed actions, the table that APPROACH uses is more like a tensor. Here, the maximum of the "can-ness" map gates the raw disparity of the can to a particular window comparator function. The output of the chosen function is an actual drive command for the robot's base.

The pixel functions used in APPROACH's space table actually consist of several processing steps amalgamated into a single transform. Consider the disparity-to-depth equation for our ranging system (figure 4-12). This function, $z(A,s)$, gives the depth of a point based on the measured position of the laser stripe in the image and the current declination of the projected plane of light (as well as the geometry of the system). We start by indexing this function on the laser plane's declination, A, which corresponds to the perceived object's y coordinate in the disparity map. This reduces the equation to a family of functions on one variable, $z_{A_0}(s)$, $z_{A_1}(s)$, $z_{A_2}(s)$, ... We then assign each pixel in the image the appropriate member of this family

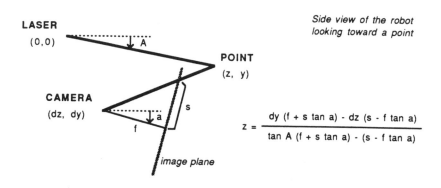

Figure 4-12. The actual depth of an imaged point can be derived from the point's measured disparity, *s*, and the declination of the laser beam, *A*.

based on its *y* coordinate. The next processing step is to check whether the robot is close enough yet by comparing the target object's computed depth to the edges of the workspace at that location. Thus, the location-based disparity-to-depth function is followed by a location-based classification function which in turn is used to choose a direction of motion for the robot's base.

$$\text{if } zA_n(s) < nearA_n \qquad \text{then advance}$$
$$\text{if } zA_n(s) > farA_n \qquad \text{then retreat}$$

Since we actually never use the real depth value for anything except this comparison, we can fold the two operations together. In fact, this is also the easiest thing to do for calibration purposes. We merely take a laser image of a plane at the near and far boundaries of the workspace and then use the raw disparity values generated as the thresholds for the comparison.

The fact that each pixel has a potentially different function suggests implementing the APPROACH module in a massively distributed fashion. Figure 4-13 shows how this could be done. At the top is the fully processed image plane in which, at each

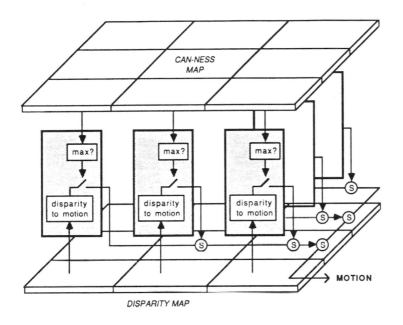

Figure 4-13. The APPROACH behavior can be implemented as massively parallel network. Each pixel checks whether a can is at its location and, if so, picks a motion appropriate to the object's depth. All the commands generated are then merged by a series of suppressor nodes.

location, the system has recorded its degree of belief that a can is there. Connected to each pixel location is a subsumption module (gray). In each module the applicability predicate checks whether its module is at the maximum of the "can-ness" map. For our system, the applicability predicates simply test whether the associated pixel has been marked as the best candidate can. The active modules then gate the result of their transfer functions to their output. In APPROACH's case, the transfer functions have access to the raw disparity map which is registered properly with respect to the "can-ness" map. The outputs of all the individual modules are combined in a large arbitration network to generate a single motor drive signal. Since the visual preprocessing we currently perform will activate only a single module, the details of this network do not matter. However, one could imagine cases where the structure of the arbitration network could produce some useful action, such as selecting the module closest to the center.

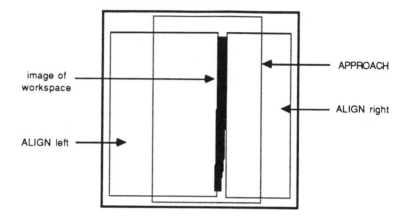

Figure 4-14. This picture shows the response regions of the modules in the control system overlaid on an actual image of the center of the arm's workspace. The robot does not approach if there is any danger that the can might move off the side of the image.

APPROACH and ALIGN form the core of the **Seek** level of competence. While their basic functions have already been explained, the robot's qualitative behavior also depends on the relative response regions of these two modules. Figure 4-14 shows a real light striper picture of the arm's workspace overlaid by the activity zones. ALIGN pays attention to the whole image except for a small vertical slice centered on the arm's workspace. APPROACH, however, only responds to cans in a small area around the workspace. If the can being pursued is near the edge of the robot's field of view, approaching may cause it to slip off the side of the image. Thus, we have adjusted the activity region for APPROACH to cause the robot to aim itself more or less toward the can before advancing.

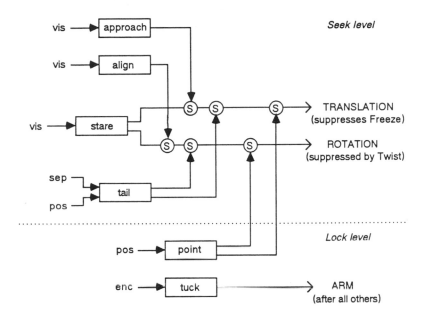

Figure 4-15. The **Seek** level pursues visually sighted cans by first aligning the robot and then approaching. To prevent damage to the robot's arm the **Lock** level prevents the arm and the base from being simultaneously active.

The other two modules in the **Seek** level (see figure 4-15) control the robot's visual attention. The lowest priority behavior is STARE actually serves two functions. If a can has been spotted recently (within the last two scans), this module grabs control of the base's motors to keep it in the same place. This lets the light striper get another look at a potentially interesting objects. Although the default behavior for the robot's base is to stay exactly where it is, this will change later when we add several levels of navigation to the system. STARE's purpose is to block any such commands which are not part of the robot's can centering routine. The other function of STARE is to abort hopeless acquisitions. If the robot has continually observed something for long time, STARE gets bored and, for a while, disregards all objects sighted. The last module in this level, TAIL, performs a similar function. In particular this module watches for the hand to be in the parked position with the fingers closed. When this first occurs, TAIL directs the robot to ballistically turn around to face the opposite direction. The purpose of this behavior is to prevent the robot from being fascinated by some object on the same table from which it just successfully removed a can.

Like STARE, the modules in the **Lock** level of the control system help coordinate the different phases of the robot's activity. For instance, we do not want the arm to be deployed until the robot has finished aligning itself with the can. Therefore we added the TUCK module which freezes the arm if the base has advanced or retreated recently. Notice that when the can is properly situated relative to the robot, APPROACH become quiescent and the base stops moving. This causes TUCK to relinquish its hold on the arm resource, effectively signalling the arm through the world that it is time for it to extend in the normal groping pattern. Much as we want the arm to remain parked when the base is moving, we also want the base to stay put when the arm is extended. For instance, if we allowed the base to rotate freely when the arm was out, the robot might inadvertently sideswipe things with the arm. This could wreak havoc with the environment and would likely damage the arm as well. To prevent this we install the POINT module which prohibits any movement of the base unless the robot's hand is at its home location.

This situation-keyed form of resource scheduling can give rise to a phenomenon akin to the ethological concept of a "displacement activity". When an animal is in a situation in which two of its innate "drives" are in conflict, it often exhibits behavior which is markedly different from that produced by either drive [Tinbergen 51]. For instance, during breeding season seagulls have territories which they assiduously defend from other birds. If another bird approaches too close, the claimant will charge forward and peck at the intruder. Yet each gull is also cognizant of the boundaries of adjacent territories. Once a gull realizes it is outside its range, it will retreats back to its home turf. The interesting situation occurs when two birds meet at a mutual boundary. Here the urge to back off and the defensive imperative are evenly balanced. What happens it is that the birds face each other and peck violently at the ground.

Our robot exhibits a similar behavior when it can not approach quite near enough to grab a can. The light striper tells the robot to advance but the robot's proximity sensors detect an obstacle ahead and enforce a stop. Neither navigation behavior directly controls the arm, however. Thus, when it detects that the robot has stopped, the arm extends and ends up pawing at empty air. One simple way to resolve this situation would be to explicitly monitor the progress of the visual approach behavior. If distance to the target does not steadily decrease, something has gone wrong and the vision system should give up. A different approach would be to loosely couple vision directly to the manipulator. For instance, we could add a module which froze the arm unless a can was sighted in the workspace. This method, unfortunately, would also require adding another special purpose interlock for releasing the arm at the deposit site.

4.4 Experiments

Here we show the performance of the can finder algorithm. Figure 4-16 shows soda cans placed in various cluttered surroundings typical of our laboratory area. The gray panels to the right are the actual light striper images. In all three cases here the robot chooses the correct object to pursue (solid black bars). Other objects which met the width and co-termination criteria but were rejected based on aspect ratio are marked with an arch-like

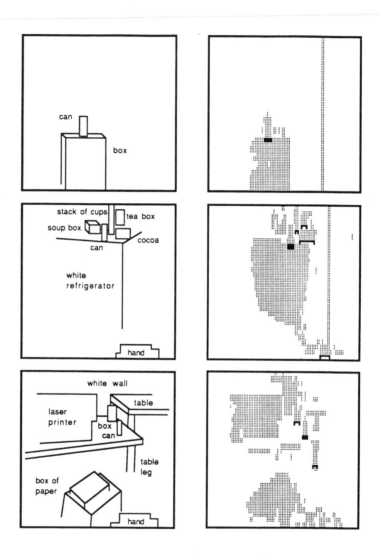

Figure 4-16. For the situations shown on the left, the actual light striper image is shown to the right in gray. The dark bar indicates the object chosen by the robot, while the arch-like marks indicate the other candidates considered.

black symbol. Notice in the third panel of figure 4-16 that the leg of the table behind the printer appears as two separate objects in the light striper image. This happens because the table leg is partially occluded by the surface supporting the printer. This is just one case in which can-like artifacts can be generated by non-can-like objects.

Figure 4-17 shows several more examples of phantom cans. In the top panel the leg of the work bench is broken into several distinct objects. In this instance, the image portions connecting the three pieces were so narrow that the front-end noise elimination step wiped them out. However, even if the robot erroneously chose one of these fragments, the semi-procedural nature of our can model would allow the robot to correct the situation. As the distance to the object decreased, the connecting bridges would appear to grow wider until, eventually, the separate segments would merge together. The longer object produced could then clearly be rejected based on its high aspect ratio. The second panel in figure 4-17 shows another mechanism by which false targets can be introduced. Here, the actual can produces a shadow on the side of the printer yielding a can-like square area to the left of the real can. This phantom will be eliminated as the robot changes its point of view, unlike a real can which remains invariant with respect to angle.

Finally, the third panel of figure 4-17 shows yet another artifact generating mechanism. Here a swath of the table top is visible at the bottom of the image. However, as the relative orientation of the table changes, the laser stripe is reflected less effectively. This leaves a raggedly outlined parabola in the image due to the streaking that results from the use of a single, highly local laser intensity threshold. A similar effect can be produced at the edge of regions by surface markings. However, as before, when the orientation of the laser system changes, so will the shapes of these regions. This is the primary advantage of using a partially procedural object model. We believe in letting the robot

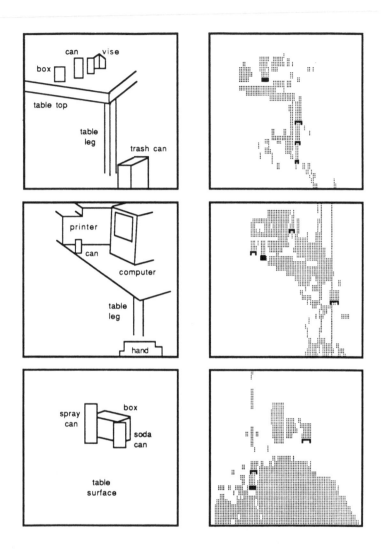

Figure 4-17. Various phenomena can give rise to can-like artifacts. The robot still manages to pick the correct object in the top two images, but fails on the third.

keep its eyes open as it moves, as opposed to a "flash bulb" approach where one image is considered sufficient for accomplishing the task. Since the robot gets many chances to correct its interpretation as the situation changes, it can afford to use simpler image processing routines (c.f. [Mysliwetz and Dickmanns 87; Horswill and Brooks 88]).

However, not only does the robot need to know what cans look like, it must also know not to respond to non-can objects. Figure 4-18 shows several scenes lacking a can. The top panel shows a person's legs; no cans were found in this image. The next panel shows a human hand in close proximity to the robot. Again, the robot is properly discriminating and does not classify the hand as a can. It does, however, mistakenly identify a background reflection as can-like. In the bottom panel of figure 4-18, we show the same image but rotated upside down. In this case the middle finger of the hand satisfies all the relevant shape constraints and is perceived as a can. It is flat and fairly narrow at the top, and has two relatively straight sides that end at the same height in the image. Without access to size information, this is probably a reasonable interpretation. Once again, after a suitable shape is identified, the robot adjusts its distance relative to the target. In this case the robot would back away from the hand to the point where the finger would fall below the minimum width threshold and disappear.

Once a can is detected, the robot moves to place the image of the can in the image of the workspace. Figure 4-19 shows the results of 10 centering trials. To obtain this data, a rigid rod was affixed to the robot's base such that its tip was directly beneath the center of the arm's 3" by 12" grab zone. As the robot swivelled to center the can, this rod's orientation and the position of its tip were recorded. The data was then transformed into the arm's moving frame of reference. From the robot's point of view, the arm does not move in order to align itself with the can; rather, the can appears to jump around with respect to

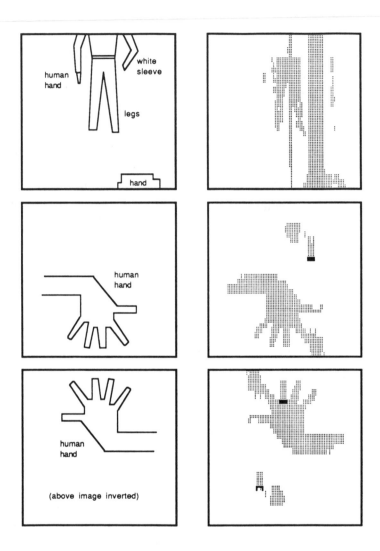

Figure 4-18. Sometimes, as in the top panel, no cans are perceived. Other times, although there are no cans in the scene, the robot is attracted to some similar feature.

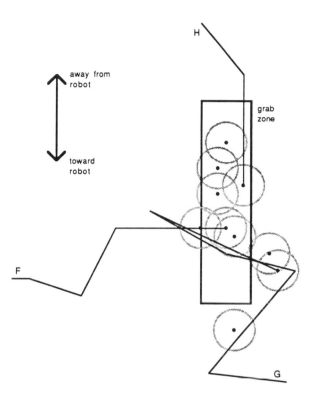

Figure 4-19. The robot is able to center cans (circles) with respect to the arm's effective workspace (rectangle). The lines show the apparent paths, relative to the arm's frame of reference, taken by several cans during the centering process.

the arm. The circles in figure 4-19 show the final location of the can while the lines show the apparent paths taken by several representative cans. The rectangle marks the limits of the robot's horizontal workspace in which the center of each can must fall. Notice that for highly eccentric cans, the robot rotates before turning (trace "F"). For cans closer to the midline the robot turns and advances simultaneously (trace "H"). Overall, the robot has a 70% success rate. Many of the errors are due to over-rotation of the base (e.g. trace "G"). This suggests implementing an alignment system with several different turn angles, thus more closely emulating a proportional controller. Big turns would be used for fast, coarse centering of the can, while smaller rotations could be employed for fine adjustment. However, even with the present "bang-bang" system, it is clear that our reduced hand-eye mapping scheme is adequate.

Chapter 5

Navigation

Using the laser rangefinder and a local groping strategy the robot is able to acquire cans that are in a 60 degree wedge in front of it. The last step is to increasing the robot's range is to move this wedge around through the environment. We do this by causing the robot to follow walls using a collection of infrared proximity sensors on its body. When, in the course of this wandering, the laser striper spots a promising candidate, it automatically stops the robot and starts positioning the arm. As soon as the alignment is satisfactory, the arm deploys itself and the normal grasping cycle ensues. Once the can is acquired, however, the robot has another duty: to bring this can back to a central collection area. For this we do not need a full-fledged map, nor do we need an understanding of how to take short-cuts or even what the quickest path back is. All the robot needs to do is to get home with its prize. This requirement is met most simply by having the robot retrace the path it took during the exploratory phase.

We have been able to further simplify our system by observing that the robot's environment is typically structured as a graph rather than as a series of wide open meadows. For instance, there are walls to follow, corridors to track along, and a warren of footpaths in what initially appears to be a room. Thus, as have others (e.g. [Kadonoff et al. 86; Wong and Payton 87]), we have broken our navigation system into a tactical component and a strategic component. The tactical part knows how to follow each of the types of paths that are found in the robot's world, while the strategic part knows when to switch between them. The strategic component's job is made much easier by using the paths intrinsic to the environment. For instance, every twist and turn of a corridor does not need to be remembered provided that the tactical component can still reliably follow it. Still, some junctions remain. Luckily, there are often local features that can be used to discriminate between directions. In fact, if we do not demand that the robot reach all parts of its environment, by always choosing the same type of branch the creature can navigate with no stored state at all.

This is the key idea of this chapter - a path does not have to be a data structure. We can instead use a simple decision procedure to choose between segments and a local navigation procedure to "record" the relevant information about each segment. Thus, our robot does not need a complete internal representation of its path. By relying on the strategic component's ability to consistent pick out the correct direction and the tactical component's stereotyped interaction with local features of the environment, the robot is able to use the world as its own representation

5.1 Sensors

The primary navigation sensors are two rings of infrared proximity detectors mounted on the torso of the robot. The sensors themselves, like the hand's crossed IRs, operate on a returned intensity basis. Two infrared emitters radiate a brief burst of photons which is bounced off obstacles in the environment and then detected by a associated phototransistor. As shown by the data in the left half of figure 5-1, the returned signal intensity falls off in a manner roughly proportional to the inverse of the distance from the sensor to the obstacle. We pass this signal through a set of 3 logarithmically spaced thresholds to yield qualitative depth information. Note that this is not an absolute measurement since the returned signal intensity depends on a number of other factors. For instance, black objects appear further away than white objects and small items are harder to detect than larger obstacles. In addition, if the reflecting surface has a matte finish (i.e. is largely Lambertian), the signal strength also exhibits a cosine dependence on the angle of this surface relative to the robot. As shown in the right half of figure 5-1, this prediction is borne out by the actual measured performance of the sensor.

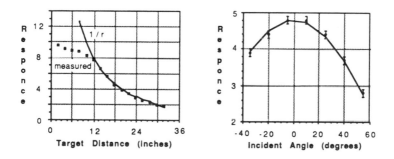

Figure 5-1. The robot navigates using 30 body-mounted infrared proximity sensors which operate in a diffuse reflective mode. The actual intensity of the returned signal depends not only on the distance to the object (left) but also its angle relative to the sensor's beam (right).

The robot has a total of 30 of these sensors. There is a ring of 16 evenly spaced sensors below the processor panels (see figure 5-2), and another ring of 14 sensors above the panels. Due to the base's mechanical linkages, both sensor rings always

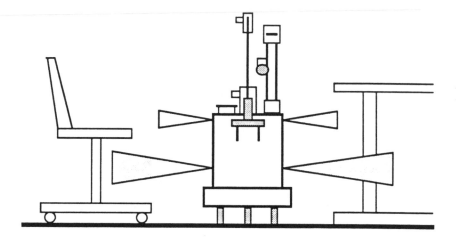

Figure 5-2. There are 30 proximity sensors mounted in two rings around the periphery of the robot. The top set sees about 1 foot while the bottom is able to sense out to 2 feet. Because these sensors have a narrow field of view, very high or very low obstructions may be missed.

remain in the same orientation relative to the robot's direction of travel. Unlike the sonar sensors typically used on mobile robots, these proximity detectors can only see a short distance. The bottom sensors typically see out to about two feet, while the top sensors are restricted to a one foot distance by mounting considerations. However, also unlike sonar sensors, there is no standoff distance - the infrared sensors can detect obstacles no matter how close they are. Our arrangement of sensors gives rise to a tapered protective shell around the robot. Any large obstacle entering this "personal space" will be detected by at least one sensor. Unfortunately, the obstacles in man-made environment are not blob-like boulders but instead have highly convoluted shapes. As shown in figure 5-2, this means objects such as table legs and the seats of chairs may actually sneak undetected past this sensory field. To correct this shortcoming, the robot continually monitors its forward progress to determine whether its motor has stalled. This allows it to detect and correct collisions with "invisible" objects.

Figure 5-3 shows the true sensory capabilities of the two rings. The dark circle is an overhead view of the robot's body and the arrow indicates its direction of travel. The plots indicate the maximum detection range for a standard 8.5" by 11" piece of white paper. The concentric circles mark the 1 foot and 2 foot perimeters. Although the on-board electronics provide 4 range bins for each sensors, we only use this outermost detection limit. We have plotted 14 readings for each set of sensors. Since the front two sensors of the top ring were blocked by the robot's hand, they were removed. Two other sensors on the bottom ring were non-operational (near the 4 o'clock position). Notice, first of all, that the ranges of the functioning sensors are not uniform - the standard deviation is almost 40%. Notice also, that even though the sensors were securely mounted, they are not perfectly aligned radially. The average deviation is on the order of 5 degrees. Finally, because the beams of the sensors are so narrow, the detection zones of adjacent sensors do not overlap. This leads to sensory dead spots between positions. Thus, when the robot rotates the proximity pattern sometimes seems to "sparkle" as skinny obstacles move into and out of the individual beams. Our control system must take account of this phenomenon, as well as the range variations and the geometric deviations from a purely radial pattern.

The robot also has onboard a compass to help it navigate. This is a Zemco digital flux gate compass which has been modified to increase its update rate. Still, it takes about a tenth of a second to settle thus making it unsuitable for use in a path segment integration scheme. Also, as shown in readings of figure 5-4, the device is not globally consistent. The circles in the diagram represent the size of the robot and the black bar inside points in the direction the robot believes to be north. The other bars indicate the directly perceived east, west, and south

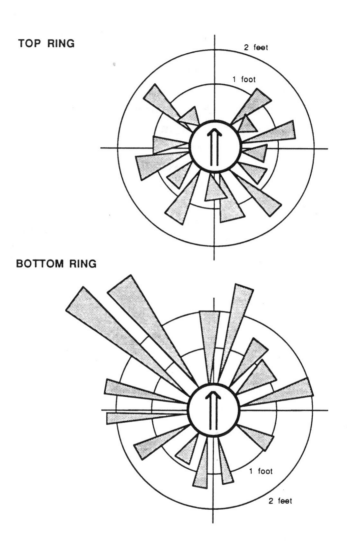

Figure 5-3. The robot's rings of proximity sensors are not very uniform. The measurements shown here indicate that the sensors are not perfectly aligned radially. They also reveal that the actual ranges of the individual sensors vary considerably.

directions, which are not always orthogonal to each other. If the robot is facing north and wants to go west, it might turn more or less than 90 degrees (the compass is not linked with the wheel encoders). In general, the readings provided by our compass are locally consistent to about 3 bits, and globally consistent to 2 bits. That is, the compass never deviates more than 45 degrees from the true direction over our test area, although the region qualitatively marked as "north" may vary in angular width. Some of the more pronounced shifts can be explained by the proximity of objects with high iron content (stippled regions in the figure). In fact, it is important that the compass be elevated from the floor (about 24" up on our robot) due to the structural steel and rebars in the concrete. Again, we must take care to design our navigation routines with these limitations in mind.

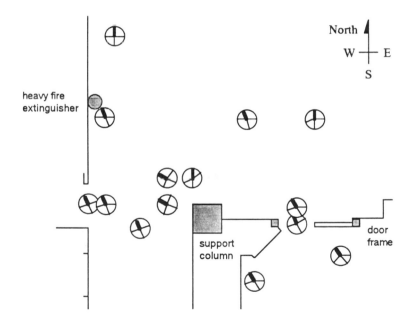

Figure 5-4. The robot also has an onboard compass for navigation. As shown in these measurements the compass is only locally consistent. The heavy line in each circle indicates the perceived direction of north, while the other lines point east, west, and south.

5.2 Tactical navigation

The robot's tactical navigation system is similar to the exploratory groping of the arm in that both follow along a surface using only local cues. However, instead of a collection of special purpose one bit sensors, the navigation system has a much richer source of information, namely the two rings of body IRs. This leads to a control structure which more closely resembles the "space tables" of the vision system than the melange of simple behaviors employed by the manipulation system. Yet, we do not simply model each possible situation and then tell the robot the appropriate action to take in each case. Rather, the proper following behavior emerges naturally as the result of two competing imperatives. First, as a basic protective measure we require that the robot not hit any obstacles and veer around obstructions in its path. We then add to this the desire to simultaneously remain within sensor range of some part of the environment, much as ancient mariners would stay within sight of land. The only way to stay near an object while moving is to follow it; thus, the robot ends up tracking along the "coast" of its world.

The actual control system for our robot is shown in figure 5-5. Only the lower 3 levels of competence are required for local navigation. The most basic level, **Stall**, causes the robot to aimlessly bumble about its world without getting stuck anywhere. This extends the robot's can collection range by moving its optical field of view around in space. The TREK module is the primary source of motion commands since it constantly urges the robot to proceed forward. If, in the course of its perambulations, the robot hits an object or runs into a wall, the STUCK module will sense that the base is exerting itself yet the robot is not moving. To correct this and disentangle the robot from the obstacle, it causes the robot to back up for a while. However, if nothing else is done, the robot will soon advance

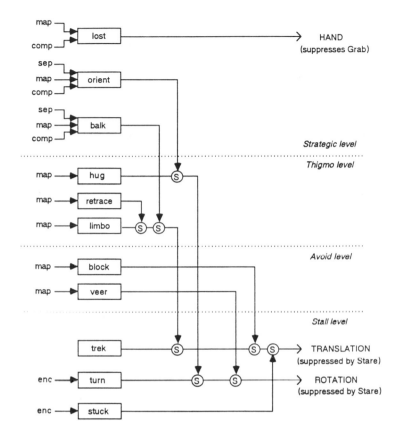

Figure 5-5. The robot's tactical navigation skills are embedded in the **Stall**, **Avoid**, and **Thigmo** levels of competence. The last level, **Strategic**, causes the robot to seek global goals.

straight into this object once again. Therefore, we add one more module, TURN, which waits for the robot to start backing up then turns it briefly to the right. Hopefully the jog this induces will permit the robot to miss the barricade on its next attempt.

The next level of competence, **Avoid**, is more interesting. The behaviors in this level use the two rings of infrared proximity detectors to assist the robot in navigating around blockages. The actual avoidance behavior is broken down into two components. BLOCK function is to determine whether the robot can safely continue forward. This behavior works on the premise that the robot cannot pass through an aperture less than 3 IR readings wide. Thus, if either of the forward two sensor positions is blocked, or both of the adjacent sensors are blocked, BLOCK stops the robot. The other component of avoidance, the VEER behavior, steers the robot in close quarters and helps it escape any deadlocks produced by BLOCK. When VEER notices that there is an obstacle in the front quadrant, it swivels the robot toward the nearest freespace. Often VEER can correct the robot's trajectory before an emergency stop is required.

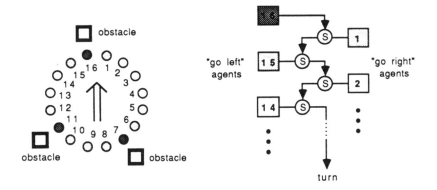

Figure 5-6. The robot grows the 3 obstacles detected into 2 forbidden regions. Each pixel in this map is then associated with a turning agent that is active (white) when the pixel is freespace. A fixed local arbitration network decides which of these agents controls the robot.

The nearest freespace is actual chosen using a "space table" similar to those in the vision-based modules. By arranging the topology of the processing elements to reflect the actual geometry of the sensors, all the necessary computation can be performed locally. VEER starts by growing all obstacles in the original IR proximity map by one sensor position on each side and then removes lone pixels of freespace. This eliminates from consideration all gaps which are too small for the robot to fit through. Then imagine, as shown in figure 5-6, that there is a special agent for each proximity pixel (this structure is "compiled out" in the actual control system). The agents on the left side of the robot output a "go left" signal when their associated pixel is sensing free-space, while those on the right generate "go right" commands. We then use a fixed priority scheme to arbitrate between adjacent and contra-lateral agents. As shown on the right in figure 5-6, agents 2 and 14 are both activated but, because of the structure of the suppression network, agent 2 wins. Thus, the robot correctly determines that turning to the right is the most efficient way to avoid the obstacle.

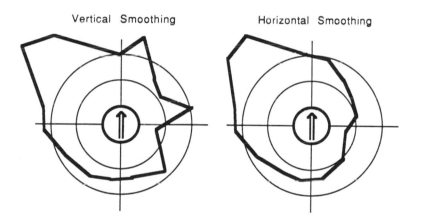

Figure 5-7. The IR sensor data is cleaned up by various forms of smoothing. In the left range diagram we pairwise OR the top and bottom readings. The right diagram shows the effect of also requiring verification by an adjacent sensor.

In addition to checking the width of apertures, both avoidance behaviors perform some essential preprocessing on the infrared sensitivity pattern. By combining the readings obtained from sensors that are aligned vertically, the top ring of sensors can fill in any obstacles missed by the bottom ring. In figure 5-7 the robot is facing in the direction of the arrow while the concentric rings mark the one and two foot offsets from it. The plot on the left shows the effective detection perimeter using the previously measured ranges (figure 5-3). While this degree of smoothing is adequate for most purposes, it makes the robot too skittish unless more processing is done for the avoidance behaviors. For instance, bright lights can falsely trigger individual sensors. In addition, as the robot moves the relative surface angles of objects change. Because of the sensors' cosine variance, this can cause a marginally reflective object to suddenly become visible or, conversely, drop out altogether. Fortunately, both of these effects are angularly localized and can be resolved by pruning proximity readings that are not corroborated by at least one adjacent sensor. As the plot on the right side of figure 5-7 shows, this additional smoothing allows a flat object, such as a wall, to approach nearer to the robot before causing a reaction.

The next level of competence, **Thigmo**, is very similar to **Avoid,** but seeks out objects instead of trying to evade them. The LIMBO behavior stops the robot if its infrared sensors detect no objects in the front half of the robot's space. The whole idea of tactical navigation is to use the constraints of the environment as paths. When the robot loses sight of the world, it becomes lost because all directions of movement are permissible. LIMBO tries to keep this from happening by stopping the robot if it is driving away from the only thing in sight. Sometimes, however, there is not adequate warning and all the IR readings suddenly vanish. This can happen if the robot is following a wall and comes to a sharply convex corner. To relocalize the robot, RETRACE backs the robot up for a while if it ever enters an open zone. This usually reestablishes contact and so gives the robot another chance to correct its heading.

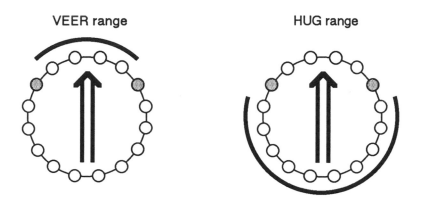

Figure 5-8. The VEER behavior tries to keep obstacles out of the robot's front quadrant. The HUG behavior maneuvers the robot to keep some thing into the front half of the robot's space. Together, these behaviors keep objects at a constant azimuth (gray sensors).

The necessary heading corrections are made by the HUG module. Like VEER it uses a space table, but turns toward the closest object sensed. HUG only generates commands when there is no object in the front portion of the robot's space. Figure 5-8 shows the active ranges for VEER and HUG assuming that there is only one object in the environment. If this object falls in the field of view of one of the front 4 sensors, VEER attempts to turn away from it. That is, VEER's applicability predicate is activated by the presence of any object in the front quadrant. This then gates the transfer function which, as described earlier, is implemented as a space table. By contrast, if the single object lies somewhere in the range of the back 10 sensors, HUG rotates the robot toward it. If the object happens to fall on one of the sensors shaded gray, neither VEER or HUG is activate and the robot proceeds straight forward. The same thing happens for objects that appear wider than one sensor reading, except that in this case it is the leading edge of the object that matters.

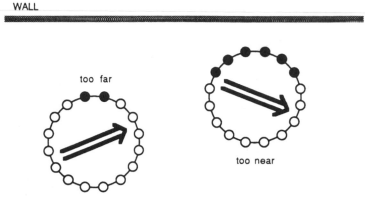

Figure 5-9. The robot turns to keep the leading edge of the sensor pattern at a certain angular position. The robot naturally spirals inward (left) if is too far from the wall, whereas it shies away (right) if it is too close.

Thus between the desire to avoid obstacles and the urge to stay near the edges of a room, a useful property emerges. As shown in figure 5-9, the interaction of these two levels of competence allow the robot to follow walls at a set distance. If the robot is too far away, HUG will steer it more toward the wall in order to bring the edge of the sensory pattern into the front 6 sensors. Similarly, if the robot is too close, VEER will reorient it outward to avoid collision. Only when the robot is parallel to the wall and at the appropriate distance are both these modules quiescent for any length of time. Thus, we can follow walls without having to make any planarity assumptions and without explicitly fitting the data to a model over time. This is especially useful for the irregular walls formed by the piles of terminals, repair carts, bicycles, and bookcases typically found in academic institutions. While similar to a wall following algorithm we have presented previously [Brooks and Connell 86], this new version does not need the complexity of polar coordinate addition. Unfortunately, because the system uses IR sensors, the actual offset distance is governed by the reflectivity of the surface. Thus, the robot will stick more tightly to a dark wall than to a light one.

5.3 Strategic navigation

Now that we can follow the pre-existing paths in the environment, namely walls and corridors, we can construct a more global navigation scheme. To complete the system, the robot must have some principled means of choosing between routes when it reaches the juncture of two or more paths. Yet we do not need a full-fledged map-based scheme that allows the robot to take novel paths through the world. The task of our robot is to simply collect cans and bring them back to some central location. As long as the robot consistently arrives at home, we are not concerned with how efficiently it does this. As we have pointed out previously [Connell 88a], it is sufficient to merely remember the outbound path and replay it in reverse to get the robot home. Once can imagine the robot, like Theseus, unrolling a spool of thread as it moves about the world, and then, once it has found a can, following this thread to eventually escape the labyrinth.

One way to accomplish this would be to record the sequence of turns made as the robot passed through doors. However, this involves the use of persistent state which might not accurately reflect the world, especially after someone closed one of the doors. Employing another analogy, the robot might instead mark its path by leaving behind a series of "bread crumbs" which pointed the way home at each intersection. Then, even if one route is temporarily blocked, the robot still has a chance of finding another marked intersection further along the path. Figure 5-10 shows a series of rooms in which all the choice points have been marked. The squares signify path junctures and the arrows specify the direction to take at each branching in order to return home. No matter where the robot ends up, if it takes the directions indicated, it will always get back to its starting point.

We are still left with the problem of associating "landmarks" with directions. If the doors are individually distinguishable (for instance, they might have numbered infrared beacons [Kadonoff, et al. 86]), the robot could build up a table linking places to routes. For instance, every time it encountered an unknown landmark it could record its direction of approach. This direction could be in specified in some global coordinate frame (i.e. a compass heading), or could be relative to some canonical orientation of the landmark itself. One could even imagine changing the pointers over time to make the robot's expeditions more efficient and to allow it to take advantage of recently discovered shortcuts. Unfortunately, unless we alter the natural environment, all doors look essentially the same to the limited sensory capabilities of our robot. Thus, it would be much better if each doorway was marked with an easily perceivable arrow pointing in the home direction. This is a technique actually used in spelunking since many of the intersections in dimly lit underground caverns are not particularly memorable.

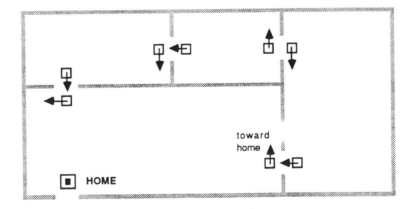

Figure 5-10. Strategic navigation can be achieved by leaving signposts in the environment. At each path juncture the robot records which direction of travel leads to the home location.

Yet laying down signs also involves modifying the environment, something we would like to avoid if possible. In addition, as Hansel and Gretel discovered, these signs might be inadvertently altered or disappear altogether over time. If each door instead had some permanent directionality already built into it, we could take advantage of this to build a more robust system. For instance, using another sensory modality such as sonar, we might be able to detect the hinge side of each doorway. Since doors are seldom opened fully, the depth would fall off sharply on one side and more slowly on the other side as it hit the angled door. We could then interpret this asymmetry as the direction to travel in order to return home. Of course, the distribution of door hinges limits the portion of the environment that the robot can safely traverse. It may be that the first doors that the robot encounters are all oriented in the wrong direction. That is, the hinges do not indicate the direction home. In this case, the robot would be stuck traversing its initial path forever.

We do not directly use any of these proposed navigation schemes. However, the real system is closely related to the intrinsic orientation scheme described above. Instead of using a feature of the door itself, we use a global orientation reference to select a path at each juncture. In other words, like a migratory bird, to get home the robot always travels south (see figure 5-11 left). This extends the robot's reliable navigation range because it is no longer limited by vagaries of the building's construction. The method is also practical for implementation purposes because the orientation functionality required is particularly well suited to the robot's onboard magnetic flux gate compass. Since we are not trying to integrate path segments, the response time and global consistency of the device are unimportant. Furthermore, because intersections tend to be orthogonal, we do not need incredible angular resolution to clearly distinguish which path to take. All that matters is that the compass be local consistent over some small area in front of the door, and that it be temporally consistent so that it points in the same direction when the robot retraces its path.

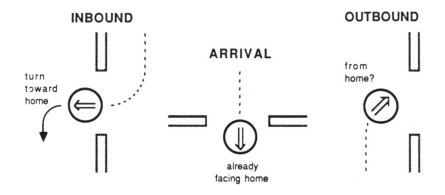

Figure 5-11. In our navigation scheme, the robot always turns south to get home (left). It has arrived at "home" when it can no longer go any further south. Thus, any east-west door is a candidate (middle). To avoid getting lost, as it explores, the robot checks its orientation at each doorway (right). Here, it has come from the correct return direction and so can continue.

This compass-directed method also has the interesting property that it automatically defines a home location for the robot. Suppose, as shown in the middle panel of figure 5-11, that on the inbound journey the robot reaches a door whose axis is oriented directly along the home vector. It passes through this door and then attempts to head in the direction of home. However, this would lead the robot out into the middle of some large open region. Since the robot is constrained to only move along walls, the LIMBO behavior halts it in its tracks. This conflict condition persists indefinitely and eventually releases the arm interlock behavior TUCK which in turn causes the robot to set down the can it has been carrying. At this point the robot becomes interested in collecting more cans and switches to the outbound mode of navigation once again. A procedural specification of the home location such as this is not incorporated in other maze solving procedures such as the "left hand" rule. Here would have to add additional sensory capabilities to detect when the robot reached the "exit" of the maze.

To ensure that the robot can always find its way home, there are a few conditions and adjustments that need to be added to the algorithm. First, if the robot ever reaches a door and finds that it has not been travelling opposite to the home vector, it should immediately turn around and head home. This situation is depicted in the right panel of figure 5-11. By terminating its wandering at this point we guarantee that the robot will never execute a path segment that it does not know how to invert. Of course, for the algorithm to work we must also assume that a new, improperly oriented door is not added to the environment between the inbound and outbound phases of one of the robot's expeditions. Another is that before heading home the robot must first turn around 180 degrees. This inverts the most recent portion of the path and causes robot to pass through previously certified doors rather than unknown ones. Fortuitously, the TAIL module in the visual servo system already performs a similar maneuver each time a can is retrieved. Since the robot is only able to detect cans roughly in line with its direction of travel, this turn will indeed reverse the robot's course.

The final condition necessary for success of the algorithm is that the robot must pass through every door it sees. This is because the robot turns toward home only as it is *leaving* a door. Suppose the robot found a door on its outbound journey and that the robot had approached it from the correct direction but chose not to go through. When it again encounters this door on the way home, since it has not just left a doorway, it is free to pass through instead. This would cause the robot to end up lost in an uncharted portion of the world, or reach a different final door which happened to be oriented in the same way as the home door. By symmetry, because the robot passed through every door it found on the outward journey, it must also pass through all doors on the inbound phase.

The complete navigation algorithm is embodied by the **Strategic** level, as shown in figure 5-5. The most basic behavior is ORIENT which causes the robot to attempt to align itself with the home vector as it passes through a door. The companion behavior BALK prevents the robot from moving unless it is roughly in line with the home vector. Both behaviors are activated by the presence of a nearby door. The robot detects such places by looking for particular features in its local IR scan. If there is both a narrow aperture (3-5 readings) and a wide aperture (> 8 readings), the robot has discovered a door. The second clause in this definition was added to prevent the robot from responding inappropriately in close quarters, such as when traversing a corridor. The commands generated by ORIENT and BALK override the **Thigmo** level of the controller but defer to the **Avoid** level. Thus, as the robot passes through the door it will repeatedly attempt to turn toward home. To allow forward progress to actually be made, the alignment threshold used by BALK must be set fairly loose (e.g. +/- 30 degrees).

Both the ORIENT and BALK modules are only active on the inbound leg of a journey. Since this usually occurs immediately after the hand has grasped a can, these two behaviors monitor the separation of the fingers to determine whether it is time to go home. However, as mentioned earlier the robot sometimes needs to retrace its steps if it gets to a door at which the home vector is pointing in the wrong direction. To accomplish this, every time the robot approaches a door the LOST behavior checks whether the robot has come from the direction of home. The tolerance on this test is set fairly broad (+/- 55 degrees) because the robot typically angles inward toward the wall before it actually detects the door. Then, instead of generating a whole new set of behaviors for this case, LOST simply simulates the environmental conditions produced in the normal course of affairs. If the robot has approached a door from the wrong direction, the LOST module forces the hand to close for a length of time.

This is like tying a string around your finger to remember something. It allows the robot to accomplish its task without requiring the addition of any explicit internal state bits. Of course, the method only works adequately when there are a small number of things to remember. However, this is the case for our navigation scheme and other tasks as well (such as groping for the can). Furthermore, the use of this **particular** piece of external state allows us to link into the pre-existing return behaviors. The closing of the hand both prompts TAIL to turn the robot around, and enables ORIENT and BALK. At each door on the way home, the monostable inside LOST is reset and the hand remains closed. Since it is only on the initial closure that the robot turns around, the homeward trajectory is not disturbed by this timeout renewal. Finally, when the robot has reached its destination and pressed the can against a supporting surface, DEPOSIT opens the fingers to release the can.

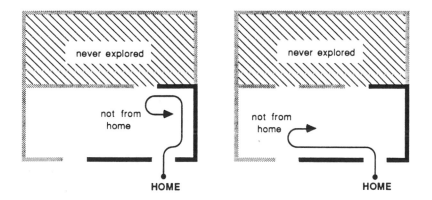

Figure 5-12. The robot can not safely travel to all parts of environments. Any door oriented the same as the home door causes the robot to stop. Thus, the top room is never explored, nor are the light colored walls in the original room.

This algorithm still has a number of shortcomings. In particular, any door oriented in the same direction as the home door blocks further exploration. For instance, the robot will never reach the upper room in figure 5-12, nor will it even investigate the gray walls in the original room. In general, the robot will only explore its environment in a direction perpendicular to the home vector. The only exceptions to this restriction are certain doors which are flush with the corner of a room. If the robot is following a wall in a direction directly away from the home vector, it will still be going in the same direction when it encounters such a door. Hence, the robot is free to pass through without fear of getting lost.

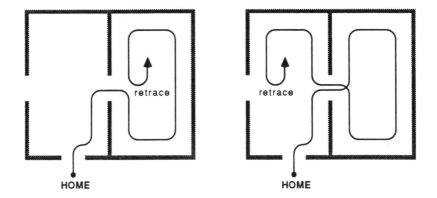

Figure 5-13. The robot cannot recognize individual doors. This can lead to inefficient homeward navigation (left) as well as highly circuitous routes (right).

The other annoying limitation of the algorithm is that the routes taken by the robot can be very circuitous. As shown on the left side of figure 5-13, the robot cannot recognize individual doors. Thus, when it encounters the door in the dividing wall for the second time all it knows is that it did not approach this door from the home direction. Therefore, instead of just passing back through the door, the robot reverses its path and again circumnavigates the room until it reaches this same door from the other side. Also, in certain circumstances as shown on the right in figure 5-13, whole extraneous loops can be added to the robot's route. Again, this is because the robot does not realize that it has passed through the same door twice and can prune its path. Nevertheless, by undertaking this arduous journey the robot does indeed get home.

5.4 Experiments

Here we investigate the actual performance of the tactical navigation system. The data shown was obtained by strapping a felt-tipped marker to the underside of the robot near its centerline. After the robot made its autonomous run, the resulting line was recorded relative to the tiling pattern of the floor. Although the pen was not always exactly centered, it was affixed to the central part of the base which, due to the synchrodrive mechanism, remains in a relatively fixed orientation with respect to the environment. Thus, the whole trace is shifted about an inch in one direction. However, the initial alignment of the robot was chosen so that this offset occurs in the principle direction travelled by the robot, so the offsets from obstacles shown in the diagrams are fairly accurate. No attempt was made to chronicle the speed of the robot, nor were the positions of pauses recorded.

The results of 5 consecutive wall following trials overlaid one on top of the other are shown in figure 5-14. Notice, first of all, that the robot's performance is very repeatable -- there are no large divergences from the basic trajectory on any of the trials. Notice also that the same wall following mechanisms works in a number of different situations. At the beginning of the course the robot is required to make a U-turn around a line of boxes. While the robot did stop briefly a few times, in no case was any backing up required nor did it hit any boxes. The next situation encountered was a concave corner such as found at the edge of a room. As soon as the robot detect the wall perpendicular to its original course it makes a detour to the right. Finally the robot enters a corridor region. Initial the hallway is wide enough that the robot cannot see the far side and therefore performs simple wall following. As the corridor narrows, the robot migrates slowly toward the centerline. This occurs because the HUG module is typically inactive in a highly constrained environment such as this.

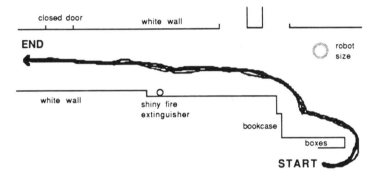

Figure 5-14. This is a composite of 5 actual runs made by our robot. Notice that the same tactical navigation routines let the robot competently handle highly convex areas (boxes), concavities (near bookcase), walls, and hallways.

Next we tested the robot's ability and propensity for passing through doors. The door used for this experiment was 26" wide, made of light colored wood, and had a tiled floor on each side. The diagonal on the left side of the picture is the door itself in the opened position. Figure 5-15 shows a composite of 5 consecutive trials using the only the tactical navigation system. From the data it can also be seen that the robot is able to successfully traverse the doorway without getting stuck. At several points the robot stopped and rotated back and forth a number of times before finding a clear direction of travel. This suggests that the dynamics of the robot are not closely enough matched to its sensory capabilities. As the robot moves, its sensors sweep over a small area during each sensing period. A close object may be smeared into two sensor readings whereas a poorly reflective object may be missed altogether. In addition, as mentioned earlier the overall IR sensing field has dead zones within it that can cause additional confusion. Yet in no case did the robot graze the door frame.

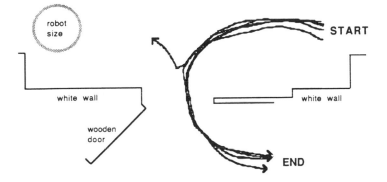

Figure 5-15. This is another composite of 5 runs of the physical robot. The same navigation skills used for wall following also allow the robot to pass safely through doors. Notice that even for this narrow door (4" clearance) the robot spontaneously goes through 80% of the time.

Notice also that the robot chose to go through the door on only 4 out of 5 trials and therefore violates one of the assumptions of the strategic navigation system. More reliable performance has been obtained with wider doors but has not been formally documented. Notice also that the robot always turns the same direction after passing through the door. This occurs because the robot starts out closer to one side and can thus sense it more consistently. Also, the door itself on the other side is not only further away but angled with respect to the robot. Nevertheless this result suggests that when running the strategic navigation layer there may be more unexplored areas than originally assumed.

Chapter 6

Discussion

We have described a complete mobile robot system for collecting soda cans in an unstructured environment. This system is unusual in that it is controlled by a collection of independent behaviors rather than a centralized program, and because it uses a minimal amount of state. Furthermore, the robot satisfactorily accomplishes its task without building complex internal maps of its environment or performing sophisticated planning operations. A number of experiments were conducted to document the robot's abilities and illuminate its limitations. Here we summarize our overall findings and suggest avenues for further research.

6.1 Spatial representation

Our robot uses a variety of spatial representations, but none of them are complete and detailed. For instance, the system incorporates information about the rough size and shape of a soda can, the sensory appearance of a wall at the correct distance for following, and the tactile signature of a supporting surface. The most prominent feature of these fragments is that they are all "task-oriented". By this, we mean that perception is directly coupled to action. Our robot does not try to pack its sensory information into some general purpose structure to be stored in a database somewhere. Instead, there is a tight coupling between particular motor acts and special purpose recognition procedures. This lets the robot concentrate on just those aspects of the environment that are relevant to its current activities. For instance, the robot does not need a deep understanding of obstacles in order to navigate around them. Contrast this to classical artificial intelligence research in which the functional and structural significance of items such as chairs has been paramount.

Our representations are not only minimal with respect to the resulting behavior, they are also distributed according to the resources required by the local situation. The job of responding to a particular stimulus is split up among a number of different modules. For instance, the robot's "knowledge" of cans is divided between the visual alignment system and the the local arm controller. The LOVPs "understand" (in a limited sense) what the outline of a can looks like and what size it should be. The OVER, EXTEND, and TWIST modules "understand" the way a can fits into the hand and how to get it there. Likewise, the perception of supporting surfaces is embodied in BOUNCE and SURFACE, and environmental obstacles are recognized collectively by BACK, UNCRASH, STALL, and COLLIDE. This adds a measure of robustness to the system. As an extreme case, consider what would happen if the crossed IR sensors fail. The hand would still descend and grope along the surface of the table, but the robot might knock over the can when it was encountered. The robot is still able to do something useful - it merely acts at a lower level of performance. Our sensor "fission" approach is to partition the desired behavior of the arm controller into separate fragments, one for each type of motion. We then analyze the dependence of each of these characteristic movements on the different sensory modalities available.

Suppose that the crossed IRs are still functional. Let us then review how the hand would be programmed to avoid local obstacles. Under our methodology we first ask what the component motions of the task are, and then what sensory information is need to recognize the associated situations. For instance, when an ungraspable object is in front of the hand we want the hand to retreat and recenter itself. The BACK module produces the appropriate arm motion and needs only the information from the crossed IRs to tell when to act. We create another module, TWIST, with similar triggering conditions to control the other actuator resource required for the overall motion (i.e. the base). Sometimes, however, the offending obstacle is invisible to the proximity sensors. To remedy this we create another module, UNCRASH, which also causes retraction, but uses only tactile information to determine when an obstacle is present. However, there are some situations which call for qualitatively different avoidance maneuvers. For instance, when the hand comes up to a vertical wall we want it to rise until it reaches the top. This is the function of the OVER module. Like BACK, it uses the crossed IR sensors, but now interprets them in a different manner.

By keeping the perceptual modalities distinct, we avoid the difficult task of cross-calibrating disparate sensors. By splitting control according to actions, we allow the prevailing environmental conditions to determine the actual trajectory taken. By separating the actuator resources, we can tune up each part of the system independently. Contrast this to a traditional, perceptually integrated approach. Here we would use the information from the tip switches, the crossed IR beams, the joint position sensors, and the base rotation encoders to build up a local model of obstacles to avoid. This model would then be used to tell the robot how to thread its way through the environment to reach its goal. Yet our distributed system performs the functions required for successful operation without first fusing the all the available sensory data. Not only is there less work involved, but since the action modules are not expecting complete models anyhow, the system is able to "improvise" with only partial information. In general, the extent of a creature's "knowledge" should be judged based on how it acts, not necessarily on how it thinks. Thus, since behavior fusion performs as well as sensor fusion for the task our robot has been given, both approaches can be considered equally competent in this case.

We can take the behavior fusion approach still further and use the effects of the behaviors themselves to pare down our representations. In earlier work [Brooks and Connell 86], we showed that we do not need a scale invariant model of doorways if the robot is guaranteed to follow walls at a particular distance. This result has been used on Herbert as well. In fact, our robot's model of its world is largely composed of behaviors. It does not record the specific path followed but instead assumes that on the way home invoking the wall following routine will yield the same result as before. Kuipers has also used a similar technique for specifying routes between distinctive locations [Kuipers and Byun 88]. Sometimes two behaviors even actively cooperate to recognize an item. For instance, although many of the physical dimensions of a can are directly embedded in the control functions, nowhere does our robot have *explicit* knowledge of these parameters. The ALIGN module simply picks a promising candidate and APPROACH drives toward it. If the object is too big, as the robot gets nearer the object's increased angular width will eventually cause ALIGN to reject it. If the object is too small, APPROACH will back the robot up until the item vanishes into obscurity. Thus, between these two behaviors the property of size emerges. However, this concept is of size is not consciously recognized or reified in anyway. This is a disadvantage if we want the robot to do any sort of introspective learning about objects.

6.2 Distributed systems

Much as processes can be used as a form of representation, they can also be used in place of actions as primitives for planning. This is how most of our system works: each behavior can be considered a process whose order of invocation is determined by the world. As was seen in the arm experiments involving pedestals and barriers, the structure of the environment effectively "programs" the robot to take an appropriate path by triggering various behaviors in a particular sequence. Thus, instead of consciously planning from a map, we use the world as its own representation and allow the situation to control the flow of events. This environmental activation property also aids the decomposition of our control system. As we add levels of competence we do not have to explicitly chain the different actions together; a new level merely puts the world into a configuration which is recognizable to a lower level. We have also taken advantage of this same effect to coordinate groups of behaviors. For instance, stopping the base causes arm extension, stopping the hand causes arm retraction, and a long pause causes the robot to turn around and head home. The ability of different processes to communicate through the world is a direct result of our data-driven procedure invocation scheme.

This lack of an explicit control structure is one consequence of our decision not to go "meta". Many systems such as GAPPS [Kaebling 88] and SOAR [Laird et al. 87] rely on a recursive control structure which may or may not eventually bottom out. Often they have an attractively simple base level in which most of the computation is supposed to occur. However, when unusual features are called for, one invokes a special purpose meta-level facility to resolve the problem. Yet frequently most of the interesting computation occurs in the unstructured meta-level rather than the elegant foundation. The other problem with such systems is the faith they place in the comprehensiveness of the meta-level. One is tempted to assume that any problem that arises can be cured by the meta-level, or that adding the proper meta-level controls will drastically improve the system's performance. On the other hand, a meta-level can be a useful tool for coordinating the functions of lower levels. In particular, it provides a medium for the cooperative resolution of command conflicts and for the long-range scheduling of resources.

In our system there is no infinite regress of meta-levels. Each module is divided into a continuous transfer function and a discrete applicability predicate. While the applicability predicate can be considered to be "meta" to the transfer function, there are no more hidden levels. Likewise, our arbitration network is a straightforward fixed priority scheme. However, this does not mean that there is no implicit <u>structure</u> to it. We have arranged things so that more specific behaviors suppress general-purpose behaviors, and so that behaviors which occur later in a typical sequence take precedence over earlier behaviors. In addition, behaviors are ordered along an immediacy spectrum. The least important are behaviors with fixed action patterns. Next come event-triggered behaviors, followed by reflexive agents, and finally the regulatory procedures at the top end. Yet this ordering hierarchy is pre-wired and static - it is not the product of some piece of dynamic meta-level machinery.

So how can we extend such a distributed system in a principled way? Theoretically higher level motor control signals could be injected anywhere in the system. These might come from a neural network trajectory generator, a classical path-planner, or a stereo vision system. In previous work [Connell 88a] we suggested that a stack be used to record the turns made by the robot. Playing back these choices straight into the robot's direction control wire would then get it home. Taken to an extreme, we could add one more module labelled HUMAN which allowed the operator to use a joystick and override the commands produced by some portion of the navigation system. A different approach, more in keeping with the indirect nature of artificial creatures, would be to selectively switch on or off whole groups of behaviors. Our robot does something like this when the arm changes over from exploration to retraction. Here PARK and PATH suppress the more general groping behaviors DESCEND, SURFACE, and EXTEND. Along similar lines, other researchers have investigated how enabling only certain navigation routines affects the path of a robot [Anderson and Donath 88a]. This approach is reminiscent of the set-point theory of trajectory generation in which all the muscles are given certain lengths and stiffnesses but the actual path is left unspecified [Hogan 84].

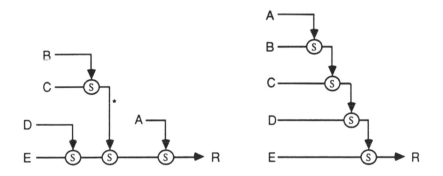

Figure 6-1. A pure suppression network specifies a total ordering on the priority of the modules so these two networks act the same. However, inhibiting the wire marked with an asterisk in the left structure has an effect which can not be duplicated in the right structure.

One way to switch on and off banks of behaviors is to make better use of inhibition nodes. Because our current arbitration scheme uses only suppressor nodes, it defines a total order on the modules. If module A suppresses module B we can say

"A > B", meaning that A always has strictly higher priority than B. Since suppression is associative, we could wire the network shown in figure 6-1 such that A > (B > C) or, conversely, so that (A > B) > C. Since these both specify the same ordering, the detailed configuration of the suppression network does not matter. However, suppose in figure 6-1a that some module inhibited the wire coming out of the suppressor node on C's output (asterisk). This would turn off both B and C but leave the rest of the modules alone. This effect can not be achieved with the network shown in figure 6-1b. Any single node that inhibited both B and C must perforce inhibit A as well. This selective excision phenomenon is potentially quite useful, especially if we group the modules in task-oriented branches. It lets us temporarily forbid the robot to attempt a certain operation (such as the one performed by modules B and C) without disturbing the rest of its control system.

However, it would be most useful to switch on and off groups of behaviors corresponding to whole levels of competence. For instance, by turning off the **Grip** level we could prevent the hand from grasping everything it touched. By turning off the **Stall** level we could keep the robot from continually wandering around. Unfortunately, the inhibition scheme above would not work for this because all the behaviors in a level are not necessarily on the same branch of the suppression network. Thus, it seems a new grouping construct and another level of arbitration might be in order. Much as the satisfaction of module's applicability predicate gates its transfer function, so might the "potentiation" of a level enable all its component modules. A similar idea is presented in a different guise in the Society of Mind architecture [Minsky 86].

To test this type of partitioned system we need to give the robot more modes of operation. A plausible extension to the current system would be to have the robot recognize more types of objects and treat them differently. For instance, there might be different "homes" for different classes of objects. Cans would go to a bin, whereas anything else went to a trash barrel. Other objects might just be collected onboard the robot and not require any relocation. For instance, the robot might drop infrared beacons at regular intervals and occasionally need to recycle some of the previously deployed units. Some objects might not have a fixed "home" at all, but require the robot to locally search for an appropriate location. For instance, if the robot found a blackboard eraser on the floor it would look around for the nearest chalk tray to return it to. This multiplicity of activities would force the robot to selectively disable certain routines to avoid resource conflicts with the proper navigation and manipulation agencies.

6.3 Limitations

The type of control system we have proposed has two main features. First, it emphasizes the use of local reactive procedures instead of detailed persistent models. This lets us sidestep many of the difficult problems involved in cross-calibrating the sensors and effectors and integrating their values over time. We pay for this with less than optimal performance. In particular, without a history of events or an overall picture of the situation, the robot faces the possibility of getting stuck in local minima or entering infinite loops instead of achieving its true goal. The second distinctive aspect of our approach is that it advocates distributed decision making versus centralized control. This, again, involves an engineering tradeoff. Hopefully, we gain extensibility and flexibility by having a number of independently combinable agents. However, we typically lose the close coordination between activities possible with a more structured system.

The strongest constraint imposed is that of temporal locality. This effectively forbids the robot to build up models from sequences of sensor readings, or to remember anything it has learned about its environment. In spite of these limitations, a mostly stateless system has the advantage that it relies solely on current, directly perceived information. Since the robot never has any preconceptions about the condition of the world, it is never fooled by a dynamically changing environment. Any robot which included a comprehensive world model would have to spend considerable effort to keep its representation accurately updated to achieve this same robustness. Yet not every aspect of the world is in constant flux; there are many useful invariants and regularities. In this sense, the totally stateless approach is too extreme. Even in our robot we have built in certain facts, such as that soda cans are oriented upright and that IR sensor 1 views a region of space directly next to that perceived by IR sensor 2. In addition, we use the state in various monostables to perform ballistic movements and to regulate the robot's mode of operation based on past events. What we have tried to avoid is storing a lot of detail about the environment. In most cases we are caching single bits of information rather than whole datastructures.

Still, there are tasks which cannot be accomplished with such limited state. The most obvious example is detour behavior involving a map. Unless the robot has an internal model of the connectedness of its environment, it cannot plan an alternative route if the usual path is blocked. The map knowledge necessary for this task is usually not built into a robot. Typically, we want to install a more general purpose navigation system and then let the robot perform its own survey of the deployment environment. However, for these tasks a full-fledged architectural blueprint of the building is not necessary. As various other researchers have pointed out (e.g. [Chatila and Laumond 85; Kuipers and Byun 88]), it is only the topological information that is crucial.

Relaxing the temporal locality condition, the robot's navigational skills could be significantly augmented with the addition of a relatively small amount of persistent state. For instance, soda cans often occur in clusters. By remembering the path it took to reach home, the robot could return to last place it found a can and thus increase its probability of finding another can. This memory might be implemented as a short list of the turns to take. To compensate for the opening or closing of doors, the robot might also remember the approximate distances or travel times between turns [Connell 88a]. However, even if this stored model should prove wrong and the robot is unable to retrace its steps, this is not a catastrophic occurrence. Because the utility of the behavior is high and the price of failure low, neglecting the possibility that the environment might change is probably an acceptable risk.

The spatial locality constraint also imposes a limitation on the robot by precluding most forms of planning. For instance, the laser striper currently withholds some potentially useful information such as the approximate depth of the can and its height above the floor. With some modifications this sensor could also be used to find the height and angle of table tops, and to form a crude three dimensional chart of the obstacles on the surface. This data could then be used to plot a trajectory for the arm so that it would go directly to the can instead of blindly groping around for it. However, we would first have to develop the transformation between image coordinates including disparity measurements and joint angles of the manipulator. This might or might not go through an common intermediate phase involving the global cartesian coordinate frame. To derive a reasonably accurate global model we would have to correct for distortion at the edge of the camera's field of view, calibrate the laser geometry to give true depth measurements, and compensate for errors in the arm servo loop caused by non-linearity in the potentiometers and static loading of the manipulator. Even once we accomplished this, the system parameters are liable to drift over time thus requiring an additional layer of adaptive control. Even more arduous calculations are required for general purpose asynchronous sensor fusion (cf. [Shafer, Stentz, and Thorpe 86]). Local algorithms circumvent most of this problem of establishing and maintaining consistency between subsystems.

In some cases, however, sensor fusion is both beneficial and easy to accomplish. Suppose we want the robot's manipulator to respond to a vertical surface in two different ways. If the robot's hand approaches a wall we would like it to descend until it finds a flat area. If the hand instead reaches a corner between the base of an object and a supporting surface, we would like it to rise above the object. No one sensor has the information to distinguish between these two situations. We need to combine the tactile information from the fingertips with the proximity readings from the forward facing IRs to generate a local spatial model. Since the two sensors are fixed relative to each other and constitute only a few bits, this merger is easily accomplished. However, the fused model need not be stored in some central database, or even be explicit in order to generate the required response. We might instead construct an ordered pair of behaviors to achieve the desired functionality. For instance, the DOWN module could drive the hand straight down every time the front IRs saw something. The other behavior, call it BOTTOM, would be able to recognize corners and would take precedence over DOWN. This module could be "primed" by setting a monostable every time the fingers touched something. Then, if the monostable was still on when the front IRs became active, BOTTOM would drive the hand upward instead. In essence, the recognition procedure for corners has been directly combined with the action procedure associated with this situation; the intermediate geometric representation has been compiled out (cf. potential fields in [Connell 87; Connell 88b]).

The other hallmark of our control system, the independence and isolation of the modules, can impair the robot's ability to coordinate its actions. One aspect of this is that there are no channels for the exchange of information. Thus, modules are forced to communicate by affecting the exterior world. One problem is that the "signals" used might also be generated spontaneously by some other activity. For instance, if the robot is waiting in line at a can collection center, it might make such slow progress that it starts extending its arm and ends up goosing the next person in line. Sometimes more subtle signals are in use and we must take care not to disturb them when adding additional behaviors. For instance, when the robot reaches a doorway it assumes that its body is aligned with the direction it came from. This would not necessarily be the case if we added an extra module which slowly rotated the robot's body as it moved in order to pan the laser striper back and forth.

These examples argue for the installation of a few specific centralized communication depots or lines such as a *retract-the-arm-now* line and a *average-recent-heading-was* line. The retraction line might be regarded as a complex effector resource, in the same way that the cartesian controller for the arm is. Similarly, the heading indicator could be viewed as a high-level sensory primitive, much as the LOVP can recognizer is. Still, employing a small number of such control wires is qualitatively different from writing the whole control system in a standard imperative computer language. We are not so much installing a foreman which micro-manages each of his employees, as we are providing simple standardized set of linguistic primitives. With this minor change, the operation of the robot could be made significantly more reliable.

Modules also "communicate" by reducing the current situation to one which some other module can cope with. There is no way for one module to directly pass digested sensory information to another. This is especially irksome if the second module in some sequence fails when the first module could have succeeded. For instance, once the IR proximity detectors on the front of the hand change from on to off, the robot has enough information about the sensed object's position to make a ballistic grab for it. Yet, instead, the EXTEND behavior attempts to trigger the beam sensor between the fingers in order to initiate grasping. However, if the finger sensor's receiver is bathed in bright sunlight, it may not respond to the can. If there were some medium for communication between the two processes, at least GRAB could try to adjust its sensing threshold to compensate for the ambient illumination. Similarly, when the robot is wandering around its environment it usually is following the edge of some physical object such as a bench. These objects are usually sites at which soda cans are found. Yet, the navigation procedure does not tell the vision system on which side of the robot cans are likely to be seen. This might make a difference if two equally plausible can shapes are seen but on opposite sides of the image. Thus, it might make sense to weaken the independence stricture in certain cases and allow behaviors to modify each other's parameters.

6.4 Extending arbitration

Our system is also limited by the method used to combine conflicting commands. As mentioned earlier, we currently use a simple priority scheme to arbitrate between proposed actions. This is a purely competitive process; in no case are two commands combined to yield a compromise. Furthermore, the relative priorities of the modules do not change over time - the suppression network cannot be dynamically reordered nor can modules tag their commands with a measure of certainty or urgency. We chose this scheme because it had a very simple semantics and proved sufficient for all the tasks we wished to accomplish. However, should the need arise, there are many ways in which it could be modified to make it more flexible.

One variation we have already explored is the use of partial suppression. Recall that we broke the hand force control bundle into two distinct wires and then modified the individual lines separately. This allowed us to use standard suppressor nodes to block action in only a certain direction. The actual encoding we used was inspired by imagining the effect of two wires innervating a pair of antagonistic muscles. If a single wire is "on", one or the other muscle contracts and the joint moves in some direction. If neither wire is "on" the joint swings freely, whereas if both are "on" the joint locks rigidly in place. Thus, the "off-off" combination is interpreted as "don't care". This encoding has the useful property that a command is not valid unless at least one of the two wires is high.

One benefit of this system is that it reduces the complexity of suppressor nodes, should we desire to build them out of discrete gates. As shown in the figure 6-2, we start by OR'ing the control wires together pairwise to guarantee that the output will be valid (at least one line will be "on") if either of the two inputs is valid. Next we must disable the inferior input (L1 and L2) if the dominant input is valid (either L2 or R2 is active). However, there is no need to disconnect L1 if L2 is active; the output of the OR gate will be the same regardless. The only other case in which the dominant module is active is when R2 is high. Thus we use this signal to block L1, and similarly use the L2 signal to block R1. This leads to the interesting observation that suppressor nodes can be built of "inhibition" units (the AND-NOT gates) and "augmentation" units (the OR gates). Thus, we might consider augmentation, rather than suppression, to be one of the basic behavior fusion primitives. This new mode of combination is particularly useful when two modules generate similar commands.

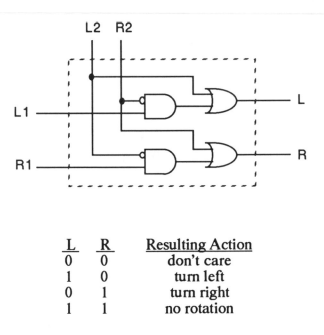

L	R	Resulting Action
0	0	don't care
1	0	turn left
0	1	turn right
1	1	no rotation

Figure 6-2. With a two wire command scheme suppressor nodes become much simpler. Each of wires in the dominant command bundle "augments" its corresponding member after first cross-inhibiting the opposite wire.

On a different track, the arbitration scheme could also be altered to allow graded responses. We usually think of the mode memory within a module as a binary storage element, yet there is no conceptual reason why it could not carry more information. For instance, a module's applicability predicate might actually be a utility function that yielded a continuous range of values. The mode memory would then act as a "peak follower" whose output would slowly ramp down after its input descended from a local maximum. Given that the output of the mode memory reflects the module's conviction that its output is appropriate, one might also modify inhibiter and suppressor nodes to act more like valves than switches. The activity level of a module would control how much it suppressed the output of another, while the relative activity levels of two modules would determine how their values were mixed in a suppressor node. However, the formulation by which such mixing should take place is unclear.

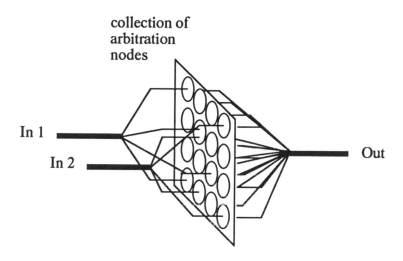

Figure 6-3. To find the equations governing graded-response arbiters we construct a surface filled with primitive nodes. Two input bundles are then mapped randomly and independently to various nodes in the cluster.

One way to derive an appropriate mixing expression is follow up on partial suppression idea and take the dissection of command bundles to an extreme. Imagine a massively parallel, statistical form of arbitration in which all transmitted values are represented in unary. Suppose we have a large collection of wires and that we activate a number of them proportional to the quantity being encoded. Now imagine two bundles of these wires impinging on a surface filled with arbitration nodes. The outputs of these nodes form the output of the whole arbitration unit. As with the inputs, the resulting value is represented by the fraction of wires that are active. To analyze this scheme we assume that the two sets of wires are assigned to nodes randomly and independently. Therefore, if v_1 is the proportion of wires from the first bundle that are active and v_2 is the number from the second bundle, the expected fraction of nodes with *both* inputs active is $v_1 v_2$.

The final outcome depends on what kind of arbitration units we use. If the surface is composed of augmentation nodes, the output value v is given by:

$$v = v_1 + v_2 - v_1 v_2$$
$$\approx v_1 + v_2$$

If v_1 and v_2 are both small, an augmentation cluster just adds its two inputs. To allow the full range of values to be used, we can create k (say 10) times more intersection points than needed and then have each output wire OR k of these points together. If the surface is instead made of inhibition nodes, we get:

$$v = v_1 - v_1 v_2$$
$$= v_1 (1 - v_2)$$

where v_1 is the value on the inferior input and v_2 is the value on the dominant input. Thus, the inferior value is attenuated by a multiplicative factor that depends on the strength of the dominant input.

To see how a graded response suppressor node responds let us reimplement the two wire turning system described earlier. Each input now becomes two bundles of wires where L and R represent the fraction of "left" and "right" wires, respectively, that are active. The direction D specified by a command is L-R; if this quantity is significantly positive the robot turns to the left while, if negative, it turns right. The quantity that determines how this command gets mixed with other commands, its "strength" S, is encoded as L+R. Figure 6-4 shows the structure

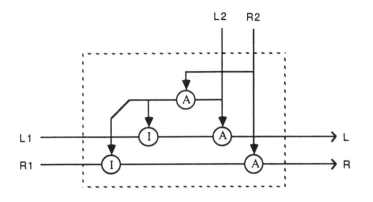

Figure 6-4. A general valve-type suppressor node for a dual bundle unary inputs.

of a suppressor node for commands of this form. Note that there are two dominant inputs to each inhibition cluster which determine how much the inferior command is inhibited. Working through the cascaded transformation we find:

$$L = L_2 + L_1 (1 - (L_1 + R_1))$$
$$R = R_2 + R_1 (1 - (L_1 + R_1))$$

By solving for the strength and direction of the resulting command we can derive a general formulation for suppressor nodes. Note that the mixing depends *only* on the strength of the dominant input (the higher subscripts).

$$D = L - R$$
$$= (L_2 - R_2) + (L_1 - R_1)(1 - (L_1 + R_1))$$
$$= D_2 + D_1 (1 - S_2)$$

$$S = L + R$$
$$= (L_2 + R_2) + (L_1 + R_1)(1 - (L_1 + R_1))$$
$$= S_2 + S_1 (1 - S_2)$$

Another approach to merging conflicting commands is to abandon the idea of suppressor and inhibiters altogether, and instead use the potential field method. The idea is to treat all commands uniformly as two dimensional polar coordinate vectors. The "strength" of a command becomes the vector's magnitude, while the "value" of the actual command translates into the vector's direction. Two commands of this form are then merged by simply adding the vectors together and finding the resultant. This method was first used in robotics by Khatib [Khatib 85] although its popularization in mobile robotics is due to Arkin [Arkin 87; Arkin, Riseman, and Hanson 87; Arkin 89].

The name "potential fields" comes from a physics analogy. We start by treating the robot as a point with a positive electrical charge and then go on to assign charges of various magnitudes and signs to other objects and locations in the world. Each of these charges exerts a force on the robot which tries to push or pull it in some particular direction. To determine which way the robot will actually go, we use superposition and sum up these force vectors. Figure 6-5 shows an example of how this method is used. We have set things up so that the robot feels a repulsion from the sides of the hallway and an attraction to a special "goal" location. This sets up an "potential" surface like a topographic map in which the walls are ridges and the goal is a pit. To find the robot's actual path from the starting location we just descend the slope by moving in the most downward direction.

Arkin uses a number of such fields to perform distributed path planning for his robot. These include radially oriented obstacle avoidance and goal attraction fields, channelled path following fields, compass-directed wandering fields, and random noise fields for escaping field plateaus. Arkin presents very detailed diagrams showing the superposition of these vector fields and gives several examples of paths travelled by the robot. Unfortunately, not all the vector fields one might want to use can be represented as strict potentials. For instance, we might want

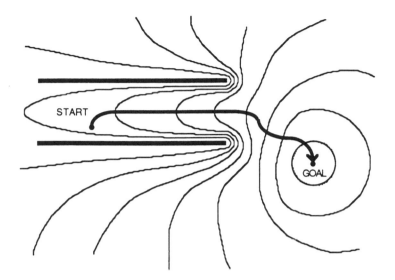

Figure 6-5. In this potential field example the robot is repelled by the walls and attracted to a particular location. The robot always moves perpendicular to the equipotential surfaces of the underlying field, shown here as contours.

the robot to proceed down a hallway in the same way that water flows through a pipe. From fluid mechanics we know that the velocity of a particle is always parallel to the surface of the pipe and varies from zero at the edges to some maximal value in the center. However, since this vector field has a non-zero curl it can not be modelled as the gradient of a scalar potential (see [Purcell 63] page 75 for other examples). This means we can not construct an energy terrain for the robot to descend. Arkin actual uses a field with non-zero curl to perform an oriented docking maneuver [Arkin 88a]. In general, dealing directly with vector fields is a more flexible approach than classical potential fields.

Even with this extension, a significant drawback to Arkin's field method is that it is world-centered not robot-centered. Since things not directly perceivable by the robot can have a non-negligible effect on its trajectory, it is necessary to have an extensive world-model. For instance, when the robot turns a corner it has to know beforehand that there will be another wall at right angles to the one it has been following. Furthermore, many times the robot is told to drive toward a goal which is not visible from all locations or, even worse, does not correspond to a landmark at all but is merely an arbitrary set of grid coordinates. In these cases it is crucial that the robot have an accurate knowledge of its position in the global coordinate frame. In addition, the vector fields used by Arkin are *totally* position-based: they ignore the robot's current heading and speed. These quantities are obviously also important for any robot which can not turn in place or which has substantial inertia. The lack of orientation also causes strange detours in the robot's path because equal attention is paid to objects behind the robot as to those in front of it. Unlike other systems (e.g [Brooks and Connell 86; Anderson and Donath 88a]), there is no way to tell the robot to go forward in the direction it is currently pointing. In recent work, however, Arkin does alter the strengths of his fields based on internal variables such as the amount of power remaining [Arkin 88b].

A still more general, and explicitly robot centered, command integration scheme is *theta-space summation* [Arbib and House 87]. This would be our choice for use in a more complex robot. Instead of having each command represent a single heading and its desirability, behaviors communicate by rating all the possible directions. This can be considered an extension of the two-wire control scheme we presented earlier, except that there is a whole range of angles (or speeds) instead of just left and right. Consider a robot approaching an extended obstacle like a fence. With the vector field approach, the robot would choose on side of the fence to go around and try to keep the trajectory of the robot close to this heading. In contrast, with theta-space the collision avoidance subsystem is allowed to have a secondary choice of direction which is not necessarily adjacent to its primary choice (see figure 6-6). Even around a particular choice the desirability distribution does not have to be symmetric: for instance, we can tell the robot that it is better to overshoot the corner of the fence than to undershoot it. As Arbib and House show, this preference pattern can be pointwise added to a similar preference pattern encoding the direction to the goal(s). The maximum of the resulting pattern then tells the creature which way to go.

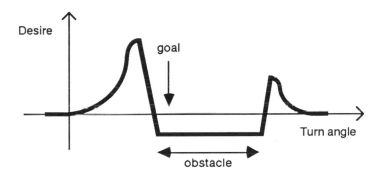

Figure 6-6. In theta-space summation each possible direction of travel is rated based on its desirability. The creature then proceeds in the direction specified by the maximum of the distribution. Here the robot must detour around a fence to get to its goal.

It can be also shown that the vector summation approach is a special case of theta-space summation. Let us represent each vector to be added as cosine distribution with the maximum centered on the desired direction and with an amplitude proportional to its magnitude. The preference pattern $D(\theta)$ for a vector at angle α and with magnitude m is:

$$D(\theta) = m \cos(\theta - \alpha)$$

As shown in figure 6-7 this is a broad, symmetric distribution. We now add together the distributions representing the two different vectors. Assuming the answer is also a cosine we find:

$$D_3(\theta) = D_1(\theta) + D_2(\theta)$$

$$m_3\cos(\theta - \alpha_3) =$$
$$m_1\cos(\theta - \alpha_1) + m_2\cos(\theta - \alpha_2)$$

$$m_3\cos(\theta)\cos(\alpha_3) + m_3\sin(\theta)\sin(\alpha_3) =$$
$$m_1\cos(\theta)\cos(\alpha_1) + m_1\sin(\theta)\sin(\alpha_1)$$
$$+ m_2\cos(\theta)\cos(\alpha_2) + m_2\sin(\theta)\sin(\alpha_2)$$

$$m_3\cos(\alpha_3)\cos(\theta) + m_3\sin(\alpha_3)\sin(\theta) =$$
$$[m_1\cos(\alpha_1) + m_2\cos(\alpha_2)]\ \cos(\theta)$$
$$+ [m_1\sin(\alpha_1) + m_2\sin(\alpha_2)]\ \sin(\theta)$$

We then match up the coefficients to yield:

$$m_3\cos(\alpha_3) = m_1\cos(\alpha_1) + m_2\cos(\alpha_2)$$

$$m_3\sin(\alpha_3) = m_1\sin(\alpha_1) + m_2\sin(\alpha_2)$$

Finally, converting to cartesian coordinates we find:

$$x_3 = x_1 + x_2$$

$$y_3 = y_1 + y_2$$

This shows that the result really is another cosine and therefore is suitable for adding to other encoded vectors. Furthermore, we can see that the components of the resultant are exactly the same as those that would be obtained with the vector summation model. Thus, the peak of the distribution is located at an angle corresponding to the resultant of the two original vectors, and the height of this peak has the correct magnitude.

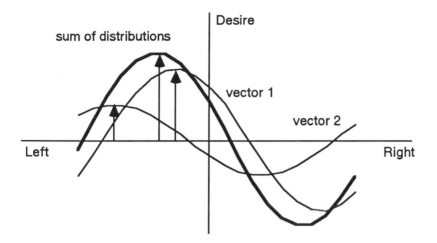

Figure 6-7. Motion commands can be represented by specifying the desirability of various turn angles. The maximum of the combined distributions is the vector sum of the two forces.

6.5 Learning

One issue studiously avoided in our current robot is learning. The robot is simply designed knowing certain fixed number of things and its performance never improves with time. There are, however, several places where adjusting parameters or remembering situations could be advantageous. For instance, the body IR sensors are notoriously unreliable. It would, therefore, be useful to include some compensatory plasticity in any module which depended on this information. Ideally, each module could discover the relevant patterns itself and adapt to any changes over time.

Yet in most cases the robot will not have someone who will lead it through a carefully chosen set of examples. The robot must be its own teacher. One interesting way to achieve this is to use the hardwired lower level routines to guide the robot's attention. For instance, another group has a robot which explores its space tactilely in an attempt to build up object models [Stansfield 88]. The interesting part of this research is that there are a variety of simple reflexes that drive the arm based on patterns of contacts. If the robot senses a flat patch, it wobbles around on the surface to determine its extent. If the finger then finds an edge, the robot carefully follows along the perceived ridge. When this linear feature ends, the robot maintains contact with the object by experimentally altering its orientation in an attempt to find another edge leading away from this corner. Most of the exploration is done by this stimulus-response system, the high level recognition routine just sits in the background and watches how the robot moves. One might consider visual routines [Ullman 83] to be in the same class. If an edge fragment is found the spotlight of attention follows along it. If a blob is detected, the system examines its boundary. If a symmetric region is found the computer inspects either end of it. These shifts can be used as the links in a graph and, when coupled with a set of local shape predicates, can be used to form symbolic data structure describing the object seen (e.g. [Connell 85; Burt 88]).

We can use the same technique to help our robot learn what it needs to survive. Take, for instance, the robot's recognition of doors. Right now this is a hard-wired ability. We manually drove the robot around to empirically determine the appropriate sensor pattern, then instilled this concept directly in the robot's control system. This door concept is primarily iconic, the only free parameter is the actual width of the door. Yet to learn anything, one must know when an instance of the goal concept, or its converse, is being presented. In this case, the robot might start wandering around and notice anytime that it makes a sharp turn. These would be positive instances of the door concept, whereas normal straight line travel would provide a background reference. Since doors are places where the robot has the choice of following several different walls, we might further bias the learning by only registering turns in which there are obstacles on both sides. This is a case where we rely on other behaviors, namely wall following, to bootstrap the learning of a new behavior.

The transfer functions used by modules could also be learned. Take, for instance, the cartesian trajectories required by our arm controller. In most cases it does not matter if the hand moves precisely up or precisely out. When the hand hits a table we only care that it has some upward velocity component. To simplify matters, one might allow the robot to use only a small set of stereotyped motions. As an extreme example, consider those trajectories in which only one joint angle changes and does this at a constant rate. The idea then is to classify the cartesian effect of such motions over the various portions of the work-space. Then, when it comes time to lift the hand, we choose that trajectory which goes closest to straight up. Suppose we start by hard-wiring in a particular single joint trajectory for each direction of motion called for. This set will typically only work for a small portion of the robot's workspace. We then start altering the direction to trajectory mapping for different regions. If a particular motion causes the finger tip switches to come on, we know the motion has a definite downward component, at least locally. This gives us the sign of the change; we can also clarify the relative magnitude of the vertical component by examining other sensors. For instance, if we want to go purely upward we would reject those motions which cause the crossed IR proximity sensors to come on. In this way, the necessary classification of motions can happen by simply watching the incoming kinesthetic and sensory patterns as the robot goes about its task.

In other cases the relevant adjustments needed are between competing behaviors. Consider calibrating the trade-off between obstacle avoidance and wall attraction. Again, this was determined by trial and error and then built into the robot. Here the appropriate error signal might be derived by watching the motion of the base itself. To properly follow a wall the robot should never have to stop because it is angled to far inward and might collide otherwise, or because it is angled too far outward and might leave the wall behind. We would start by priming the sensitivity range of each module with the minimum acceptable area. VEER should always respond to obstacles in the front quadrant and THIGMO should never let the front half space become vacant. Anytime the robot stops, the currently active module should consider incrementally extending its range while the inactive one should contract its zone of influence. One could imagine a variation of this algorithm in which the two sides of the robot were treated independently to allow asymmetric ranges to be established.

6.6 Conclusion

In this report we have described in detail the construction of a mobile robot which autonomously collects soda cans. This serves as an existence proof that a moderately complicated system can be built which is decentralized and does not contain explicit world models. The reasons for these design choices were discussed and their implications were examined in terms of the actual robot. In the course of this project, several interesting epiphenomena were also observed. These included the emergence of new ensemble behaviors, semi-procedural models for the recognition of relevant items, and the automatic enlistment of appropriate behaviors through environmental signalling. The robot described here is just the first step on the long road toward competent autonomous robots. As we have noted, a number of extensions could be made to the current system to enhance its performance and to help the architecture scale to larger systems.

Bibliography

[Anderson and Donath 88a] Tracy L. Anderson and Max Donath, "Synthesis of Reflexive Behaviors for a Mobile Robot Based Upon a Stimulus-Response Paradigm", *Proceedings of the 1988 SPIE Conference on Mobile Robots*, 370-382.

[Anderson and Donath 88b] Tracy L. Anderson and Max Donath, "A Computational Structure for Enforcing Reactive Behavior in a Mobile Robot", *Proceedings of the 1988 SPIE Conference on Mobile Robots*, 198-211.

[Arbib 81] Michael A. Arbib, "Perceptual Structures and Distributed Motor Control", in *The Handbook of Physiology: Volume 3 - The Nervous System*, Vernon B. Brooks (ed.), The American Physiological Society, Bethesda MD, 1449-1465.

[Arbib and House 87] Michael A. Arbib and Donald H. House, "Depth and Detours: An Essay on Visually Guided Behavior", in *Vision, Brain, and Cooperative Computation*, M. Arbib and A. Hanson (eds.), MIT Press, Cambridge MA.

[Arkin 87] Ronald C. Arkin, "Motor Schema Based Navigation for a Mobile Robot: An Approach to Programming by Behavior", *Proceedings of the 1987 IEEE International Conference on Robotics and Automation*, 264-271.

[Arkin 88a] Ronald C. Arkin, "Intelligent Mobile Robots in the Workplace: Leaving the Guide Behind", *Proceedings of the First International Conference on Industrial Applications of Artificial Intelligence and Expert Systems*, Tullahoma TN (also Georgia Institute of Technology ICS-88/08).

165

[Arkin 88b] Ronald C. Arkin, "Homeostatic Control for a Mobile Robot: Dynamic Replanning in Hazardous Environments", *Proceedings of the 1988 SPIE Conference on Mobile Robots*, 407-413.

[Arkin 89] Ronald C. Arkin, "Motor Schema-Based Mobile Robot Navigation", *International Journal of Robotics Research* (to appear).

[Arkin, Riseman, and Hanson 87] Ronald C. Arkin, Edward Riseman and Allen Hanson, "Visual Strategies for Mobile Robot Navigation", *Proceedings of the IEEE Workshop on Computer Vision,* Miami FL.

[Ayache and Faugeras 87] Nicholas Ayache and Olivier D. Faugeras, "Building a Consistent 3D Representation of a Mobile Robot's Environment by Combining Multiple Stereo Views", *Proceedings of IJCAI-87*, 808-810.

[Balek and Kelley 85] D. J. Balek and R. B. Kelley, "Using Gripper Mounted Infrared Proximity Sensors for Robot Feedback Control", *Proceedings of the 1985 IEEE International Conference on Robotics and Automation*, 282-287.

[Bekey and Tomovic 86] George A. Bekey and Rajko Tomovic, "Robot Control by Reflex Actions", *Proceedings of the 1986 IEEE International Conference on Robotics and Automation*, 240-247.

[Brooks 86] Rodney Brooks, "A Layered Intelligent Control System for a Mobile Robot", *IEEE Journal Robotics and Automation,* RA-2, April, 14-23.

[Brooks 89] Rodney Brooks, "A Robot That Walks: Emergent Behaviors from a Carefully Evolved Network", *Neural Computation*, 1(2), 253-262 (also MIT AIM-1091).

[Brooks and Connell 86] Rodney Brooks and Jonathan Connell, "Asynchronous Distributed Control System for a Mobile Robot", *Proceedings of the 1986 SPIE Conference on Mobile Robots*, 77-84.

[Brooks, Flynn, and Marill 87] Rodney Brooks, Anita Flynn, and Thomas Marill, *Self-Calibration of Motion and Vision for Mobile Robot Navigation*, AIM-984, MIT AI Lab, Cambridge MA.

[Burt 88] Peter J. Burt, "Algorithms and Architectures for Smart Sensing", *Proceedings of the DARPA Image Understanding Workshop*, 139-153.

[Chatila and Laumond 85] Raja Chatila and Jean-Paul Laumond, "Position Referencing and Consistent World Modeling for Mobile Robots", *Proceedings of the 1985 IEEE Conference on Robotics and Automation*, 138-145.

[Connell 85] Jonathan Connell, *Learning Shape Descriptions: Generating and Generalizing Models of Visual Objects*, TR-853, MIT AI Lab, Cambridge MA.

[Connell 87] Jonathan Connell, "Creature Design with the Subsumption Architecture", *Proceedings of IJCAI-87*, 1124-1126.

[Connell 88a] Jonathan Connell, "Navigation by Path Remembering", *Proceedings of the 1988 SPIE Conference on Mobile Robots*, 383-390.

[Connell 88b] Jonathan Connell, "The OMNI Photovore", *OMNI*, 11(1), October 1988, 201-212.

[Connell 89] Jonathan Connell, "A Behavior-Based Arm Controller", *IEEE Journal of Robotics and Automation*, 5(6), December 1989, 784-791 (also MIT AIM-1025).

[Echigo and Yachida 85] Tomio Echigo and Masahiko Yachida, "A Fast Method for Extraction of 3-D Information Using Multiple Stripes and Two Cameras", *Proceedings of IJCAI-85*, 1127-1130.

[Erdmann 85] Michael Erdmann, "Using Backprojections for Fine Motion Planning with Uncertainty", *Proceedings of the 1985 IEEE International Conference on Robotics and Automation*, 549-554.

[Essenmacher et al. 88] T. J. Essenmacher, V. G. Grafe. G. S. Davidson, and M. M. Moya, "A Structured Lighting Vision System for Dynamic Obstacle Avoidance with a Mobile Robot", *Proceedings of the 1988 SPIE Conference on Mobile Robots*, 357-367.

[Fraenkel 80] Gottfried Fraenkel, "On Geotaxis and Phototaxis in Littorina", in *The Organization of Action: A New Synthesis*, C. R. Gallistel (ed.), Lawrence Erlbaum.

[Grimson and Lozano-Pérez 84] W. Eric L. Grimson and Tomás Lozano-Pérez, "Model-based Recognition and Localization from Sparse Range or Tactile Data", *International Journal of Robotics Research*, 3(3), 3-35.

[Hirose et al. 85] Shigeo Hirose, Tomoyuki Masui, Hidekazu Kikiuchi, Yasushi Fukada, and Yoji Umetani, "Titan III: A Quadraped Walking Vehicle", in *Robotics Research: The Second Annual Symposium*, Hideo Hanafusa and and Hirochika Inoue (eds.), MIT Press, Cambridge MA.

[Hogan 84] Neville Hogan, "Adaptive Control of Mechanical Impedance by Coactivation of Antagonist Muscles", *IEEE Transactions on Automatic Control*, Vol. AC-29, No. 8, 681-690.

[Horswill and Brooks 88] Ian Horswill and Rodney Brooks, "Situated Vision in a Dynamic World: Chasing Objects", *Proceedings of AAAI-88,* 796-800.

[Ikeuchi et al. 86] Katsushi Ikeuchi, H. Keith Nishihara, Berthold K. P. Horn, Patrick Sobalvarro, and Shigema Nagata, "Determining Grasp Configuration Using Photometric Stereo and the PRISM Binocular Stereo System", *International Journal of Robotics Research,* 5(1), 46-65.

[Kadonoff et al. 86] Mark Kadonoff, Faycal Benayad-Cherif, Austin Franklin, James Maddox, Lon Muller, and Hans Moravec, "Arbitration of Multiple Control Strategies for Mobile Robots", *Proceedings of the 1986 SPIE Conference on Mobile Robots,* 90-98.

[Kaebling 87] Leslie Pack Kaebling, "An Architecture for Intelligent Reactive Systems", in *Reasoning About Plans and Actions,* P. Georgeff and A. Lansky (eds.), Morgan Kauffman.

[Kaebling 88] Leslie Pack Kaebling, "Goals as Parallel Program Specifications", *Proceedings of AAAI-88,* 60-65.

[Khatib 85] Oussama Khatib, "Real-time Obstacle Avoidance for Manipulators and Mobile Robots", *Proceedings of the 1985 IEEE International Conference on Robotics and Automation,* 500-505.

[Kuipers and Byun 88] Benjamin Kuipers and Yung-Tai Byun, "A Robust, Qualitative Method for Robot Spatial Learning", *Proceedings of AAAI-88,* 774-779.

[Kuperstein 87] Michael Kuperstein, "Adaptive Visual-Motor Coordination in Multijoint Robots using Parallel Architecture", *Proceedings of the 1987 IEEE International Conference on Robotics and Automation,* 1595-1602.

[Laird et al. 87] John Laird, Allen Newell, and Paul Rosenbloom, "SOAR: An Architecture for General Intelligence", *Artificial Intelligence,* 33(1), 1-64.

[Lewis and Johnston 77] R. A. Lewis and A. R. Johnston, "A Scanning Laser Rangefinder for a Robotic Vehicle", *Proceedings of IJCAI-77,* 762-768.

[Lozano-Pérez 86] Tomás Lozano-Pérez, *A Simple Motion Planning Algorithm for General Manipulators,* AIM-896, MIT AI Lab, Cambridge MA.

[Lozano-Pérez, Grimson, and White 87] Tomás Lozano-Pérez, W. Eric L. Grimson, and Steven White, "Finding Cylinders in Range Data", *Proceedings of the 1987 IEEE International Conference on Robotics and Automation,* 202-207.

[Lozano-Pérez, Mason, and Taylor 84] Tomás Lozano-Pérez, Matthew T. Mason, and Russell H. Taylor, "Automatic Synthesis of Fine-Motion Strategies for Robots", *International Journal of Robotics Research*, 3(1), 3-24.

[Marr 82] David Marr, *Vision*, W. H. Freeman.

[Miller and Wagner 87] Gabriel Miller and Eric Wagner, "An Optical Rangefinder for Autonomous Cart Navigation", *Proceedings of the 1987 SPIE Conference on Mobile Robots*, 132-144.

[Minsky 80] Marvin Minsky, "K-lines: A Theory of Memory", *Cognitive Science*, 4(2), 117-133 (also MIT AIM-516).

[Minsky 86] Marvin Minsky, *The Society of Mind*, Simon and Schuster.

[Moravec and Elfes 85] Hans P. Moravec and Alberto Elfes, "High Resolution Maps from Wide Angle Sonar", *Proceedings of the 1985 IEEE International Conference on Robotics and Automation*, 116-121.

[Mysliwetz and Dickmanns 87] Birger D. Mysliwetz and E. D. Dickmanns, "Distributed Scene Analysis for Autonomous Road Vehicle Guidance", *Proceedings of the 1987 SPIE Conference on Mobile Robots*, 72-79.

[Payton 86] David Payton, "An Architecture for Reflexive Autonomous Vehicle Control", *Proceedings of the 1986 IEEE International Conference on Robotics and Automation*, 1838-1845.

[Pipitone and Marshall 83] Frank J. Pipitone and Thomas G. Marshall, "A Wide-field Scanning Triangulation Rangefinder for Machine Vision", *International Journal of Robotics Research*, 2(1), 39-49.

[Popplestone et al. 75] R. J. Popplestone, C. M. Brown, A. P. Ambler, and G. F. Crawford, "Forming Models of Plane-and-Cylinder Faceted Bodies from Light Stripes", *Proceedings of IJCAI-75*, 664-668.

[Porrill 88] John Porrill, "Optimal Combination and Constraints for Geometrical Sensor Data", *International Journal of Robotics Research*, 7(6), 66-77.

[Purcell 63] Edward W. Purcell, *Electricity and Magnetism*, McGraw Hill.

[Raibert 86] Marc Raibert, *Legged Robots that Balance*, MIT Press, Cambridge MA.

[Shafer, Stentz, and Thorpe 86] S. Shafer, A. Stentz, and C. Thorpe, "An Architecture for Sensor Fusion in a Mobile Robot", *Proceedings of the 1986 IEEE International Conference on Robotics and Automation*, 2002-2011.

[Simon 69] Herbert A. Simon, *The Sciences of the Artificial*, MIT Press, Cambridge MA.

[Stansfield 88] S. A. Stansfield, "A Robotic Perceptual System Utilizing Passive Vision and Active Touch", *International Journal of Robotics Research*, 7(6).

[Tinbergen 51] Niko Tinbergen, "Chapter 5: An Attempt at a Synthesis", *The Study of Instinct*, Oxford University Press.

[Vuylsteke and Oosterlinck 86] P. Vuylsteke and A. Oosterlinck, "3-D Perception with a Single Binary Coded Illumination Pattern", *Proceedings of the 1986 SPIE Conference on Optics, Illumination, and Image Sensing for Machine Vision*, 195-202.

[Wehner 87] Rüdiger Wehner, "'Matched Filters' - Neural Models of the External World", *Journal of Comparative Physiology*, 161:511-531.

[Winston et al. 84] Patrick H. Winston, Thomas O. Binford, Boris Katz, and Michael Lowry, "Learning Physical Models from Functional Definitions, Examples, and Precedents", *Robotics Research*, Michael Brady and Richard Paul (eds.), MIT Press, Cambridge MA.

[Wong and Payton 87] Vincent Wong and David Payton, "Goal-Oriented Obstacle Avoidance Through Behavior Selection", *Proceedings of the 1987 SPIE Conference on Mobile Robots*, 2-10.

Index

I

improvisation 75
incremental systems 11, 26, 36
independence 12, 27, 56
inefficiencies of path 132
infrared sensors
  between fingers 47
  on body 112
  on hand 49
inhibition 21, 33, 154
initiation predicate 24
interlocks for arm and base 102
intrinsic directionality 127

L

landmarks 126
laser for light-striper 81
layers of behaviors 26
learning 161
levels of competence 36, 51
light striper
  equation 98
  interpretation 91
  processors 84
  structure 82
  theory 83
local minima 43
locality 12, 145
localization of cans 79
loops, problems with 43, 72
LOVPs 84

M

mediators 38
meta-level control 141
Minnesota robot 39
mode memory 24

modelling
  cans 51, 57, 93
  doors 130
  surfaces 59
  walls 124
modules, behavioral 51
monostables 25, 52
mutants 33

N

navigation scheme 128

P

packet bug 28
packets of information 20
parallel control paths 5
parallelogram arm linkages 44
partial suppression 54, 151
path, retracing of 111, 125
pedestal experiment 75
performance
  arm trajectories 74
  can finder 103
  reorientation 73
  tactical navigation 133
  visual positioning 107
personal space of robot 114
pipelined memory 86
planning
  compiled in 77
  contained in packets 28
  improvisational 75
  problems with 5, 146
plots
  body sensors 116
  compass data 117
  filter shift 84
  hand sensors 50
  proximity response 113
  workspace image 100

Perspectives in Artifical Intelligence